U0176182

观沧海

海错食单

陈橙
刘娇
陆逸
|
编译

袁小真
|
绘

【汉英对照】

The
Stories
of
the
Chinese
Seafood

中央编译出版社
CCTP Central Compilation & Translation Press

图书在版编目（CIP）数据

海错食单：汉英对照 / 陈橙编译；刘娇，陆逸编译 .
-- 北京：中央编译出版社，2020.1
ISBN 978—7—5117—3769—4

I.①海…

II.①陈… ②刘… ③陆…

III.海鲜菜肴 - 饮食 - 文化 - 中国 - 通俗读物 - 汉、英

IV.① TS971.2-49

中国版本图书馆 CIP 数据核字 (2019) 第 281435 号

海错食单：汉英对照

出 版 人：葛海彦
出版统筹：贾宇琰
责任编辑：朱瑞雪
责任印制：刘　慧
出版发行：中央编译出版社
地　　址：北京西城区车公庄大街乙 5 号鸿儒大厦 B 座 (100044)
电　　话：(010) 52612345（总编室）　(010) 52612335（编辑室）
　　　　　(010) 52612316（发行部）　(010) 52612346（馆配部）
传　　真：(010) 66515838
经　　销：全国新华书店
印　　刷：北京文昌阁彩色印刷有限责任公司
开　　本：880 毫米 ×1230 毫米　1/32
字　　数：120 千字
印　　张：6.75
版　　次：2020 年 1 月第 1 版
印　　次：2020 年 1 月第 1 次印刷
定　　价：58.00 元

网　　址：www.cctphome.com　　邮　　箱：cctp@cctphome.com
新浪微博：@ 中央编译出版社　　微　　信：中央编译出版社（ID: cctphome）
淘宝店铺：中央编译出版社直销店（http://shop108367160.taobao.com）(010) 55626985
本社常年法律顾问：北京市吴栾赵阎律师事务所律师　闫军　梁勤
凡有印装质量问题，本社负责调换，电话：(010) 55626985

千古食趣
八方美食

四时更替
温暖手绘

百味人生
不止故事

目
录

开拓，辽东的独创

第一章

Chapter

One

Liaodong Cuisine:
Pioneering Originality

　　毫无疑问，中国是世界上农耕文明的典型代表，但同时我们也不该忘记，中国人所居住的陆地，东南两面临海，拥有超过一万八千公里的海岸线。数千年来，中华儿女们一只脚踏在坚实的土地之上，另一只脚则立于浩瀚的海洋之上，在创造灿烂农耕文化的同时，也缔造了独特的海洋文化。

　　文明与历史是包罗万象的词汇，对于每个普通的人来说，走得更近一些，可能就是生活的林林总总，譬如一日三餐，而我们要讲的就是中华文明中海洋与食物的故事。

　　我们的先民从海洋中获取食物的历史十分悠久。成书于先秦的《尚书·禹贡》中曾记载："厥贡盐缔，海物惟错"。后人取"错杂非一种"之意，从此称海味为海错，在漫长的历史中又逐渐改称为海鲜，即今天的各种海产食物。以科学的眼光来看，较为温暖的海水中海鲜品类更为丰富，但这并不意味着海鲜是南方的专属。从地图上看去，辽东半岛西邻

China is undoubtedly a typical example of agricultural civilization in the world. However, it should be borne in mind that the land belonging to the Chinese spans over more than 18,000 kilometers of coastline in the east and south. For thousands of years, Chinese descendants have lived on the solid ground on one hand and have kept exploring the vast ocean on the other. Through their hard work, the people have developed a splendid agricultural culture and a unique ocean culture.

"Civilization" and "History", as two all-embracing concepts, in general, they include all aspects of life, such as daily meals. What we are going to talk about is the story of seafood in Chinese civilization.

Our ancestors have long history of obtaining food from the ocean. A master piece of the Pre-Qin literature *The Book of Documents Yugong* once recorded that: "Tributes here are mostly salt and fine cambric as well as Hai Cuo." In Chinese, "*Cuo*" means "a Variety" and "*Hai*" means "Seafood." There—fore, seafood. was referred to as "Hai Cuo (various seafood)" by the descendants. Today, it is called "Hai Xian." Scientifically speaking, a wide variety of seafood exists in warm waters ,but it doesn't mean the southern are as have a monopoly of seafood. The Bohai Bay Fishing Ground, one of China's four major fishing grounds, lies in the Liaodong Peninsular which covers the most northern coastline of China. As an important part of the Bohai Sea, Liaodong Gulf has long enjoyed a good reputation for its abundant marine products including abalones, sea cucumbers, scallops, abalones prawns, oysters, and saxidomus purpuratus.

渤海，东临黄海，占据了中国漫长的海岸线中最北端的一部分。这里有着中国四大渔场之一的渤海湾渔场，而辽东湾是其重要组成部分。辽东湾盛产海参、扇贝、鲍鱼、对虾、牡蛎、紫石房蛤等海产品，久负盛名。

对于辽东半岛依海而居的人们来说，海鲜是解决温饱的食物，更是追求美好生活的保障。历史长河中的先民们获取海鲜依靠的是较为简陋的工具，而在机械化捕捞盛行的今天，近海渔业资源面临枯竭，人工养殖成为主要方式。虽然使用了科学方法进行养殖，但捕捞的辛苦并未完全消失。海参、鲍鱼等仍需人们穿着笨重的设备深潜获取。没有养殖渔场的渔民们则需乘渔船到更远的海域进行捕捞。在辛苦劳作的过程中，各类海鲜小食成为化解疲劳的神奇良方。在锦州就有"虾油小菜"的特色食物。小虾入盐发酵，制成虾酱，在这一过程中，虾酱静置，可在酱缸表层获得珍贵的虾油。在虾油中放入豆豉、油椒，再配以小黄瓜、芹菜、芸豆、地梨等，加以调配就可制成"虾油小菜"。蔬菜的脆爽、芸豆的绵软、虾油的鲜香，交汇融合，成就了这独特的佐餐小菜。

据记载，清康熙二十一年九月，康熙皇帝到沈阳祭祖，路经广宁镇，锦州府尹特将虾油小菜送到广宁镇行宫南面框观亭，康熙品尝后连声赞好，自此虾油小菜取代了贡虾，成

For the residents of Liaodong seaside region, seafood represents hope for a better life, not just food to curb hunger. The ancestors relied on relatively simple tools to get aquatic food ,while today, inshore fishery resources are running dry due to mechanized overfishing, which made artificial cultivation the main source of seafood. On the other hand, in spite of the fact that the cultivation performed with scientific methods, the suffering of fishermen does not disappear. Wearing cumbersome equipment, fishermen need to dive deep to harvest precious seafood such as sea cucumber and abalone. Some fishermen without fish farms are even forced to go further out to the sea to fish. During the process, a variety of seafood snacks such as assorted vegetables preserved with shrimp sauce, Jinzhou specialty, becomes a magic cure for fatigue. First, enough salt is added to shrimp to obtain shrimp paste. The paste is kept for a while until precious shrimp oil floats on the top of the sauce jar. Fermented beans, oil pepper, cucumber, celery, kidney bean, and water chestnut are put into the shrimp oil. Finally, the dish is completed after stirring all ingredients for a while. This side dish brings a captivating taste experience thanks to its softness and freshness.

According to the record, in Sep., AD1682, the Qing-Dynasty Emperor Kangxi passed Guangning town on his way to Shenyang city to worship the ancestors. The magistrate of Jinzhou city managed to dedicate this side dish to the Kuangguan Pavilion, standing at the south of the Emperor Palace in the town,where the Emperor dwelled temporarily. After tasting the delicacy, Kangxi praised it so much. From that time on, the dish replaced shrimps as

了锦州的贡品。

人们在劳作过程中往往需要的只是简单的餐食，而当带着收获归家时，则会犒劳自己一番。辽东人在犒劳自己时更倾心于厚脂重盐。在盘锦，实现这一目的只需一盘油爆香螺。盘锦香螺贝壳圆胖而厚重，整体呈长双锥形，颜色多为棕色，似绒布一般，内里肉质肥美。洗净香螺，将蒜末、香菇丁等调料放油锅中爆香，加入香螺翻炒出香味，再加入料酒、白糖、酱油、盐等，然后加清汤焖稍许，撒上葱花香菜即可。一盘油爆香螺上桌，吃起来鲜香，有嚼劲，回味无穷。

在丹东，人们选择的则是更具特色的黄蚬炒米叉子。米叉子实际上是由满族食品"酸汤子"演变而来的。这一食物的做法在《奉天通志·礼俗三·饮食》中有详细的记载。但"酸汤子"酸味较大，又做法繁复，于是丹东人加以改良简化，以玉米为原料，发酵后磨成水面，经沉淀，手工或用模具制成筷头粗细的光滑条状，就做成了米叉子。米叉子富含粗纤维、蛋白质等，可炒可煮，是丹东独有的食品。金秋时节，丹东沿海特产大黄蚬子最为肥美，取黄蚬子肉，配以少许韭黄，与用凉水拔过酸气的米叉子合炒，就是当地人钟爱的黄蚬炒米叉子。米叉子的筋道、嫩爽，黄蚬肉的肥美，一起在唇齿之间鲜活起来。

the new tribute of Jinzhou city.

After hard work, people often give themselves a big meal when returning home with harvest rather than simple food during the hard work. The delicious stir-fried neptunea is an optimal choice for people in Panjin city who prefer strong flavors. The shell of the neptunea in Panjin is plum and thick, long and bi-conical in shape, and brown in colour. It is just like a flannelette. The meat inside tastes tender and delicious. What we need to cook stir-fried neptunea, wash the neptunea, saute minced garlic and diced mushrooms, then add rinsed neptunea. When there is a savoury smell add cooking wine, sugar, soy sauce, salt and clear soup, then stew it for a while. Finally, sprinkle chopped chives and coriander on it before enjoying the chewy and fresh taste which will linger long in your mouth.

In Dandong city, people opt for the more distinctive sauteed Michazi (sour maize meal) with yellow clam. "Michazi" originated from the Manchu dish Suantangzi(sour corn noodle) and its recipe is found in the *General Annals of Feng Tian: Etiquette and Custom Three.Diet*. Since Suantangzi tastes too sour and seems complicated to cook, people in Dandong then have tried to simplify and improve the meal. Corn is the raw material. After fermentation, corn is ground into batter. Then the batter needs to undergo a full sedimentation process. After sedimentation, the batter is made into smooth strips as thin as chopsticks by hand or by mold. That is often referred to as Michazi. Whether it is sauteed or boiled, Michazi is rich in crude fiber, protein, etc. In autumn, the most attractive specialty on the coast of Dandong is supposed to be yellow clams, the meat of which

is the fattest. After removing the sourness of Michazi, it is fried with the meat of the yellow clam and a few chives. Thanks to its chewiness and tenderness as well as the fatness of the yellow clam, sauteed Michazi with yellow clam has become the locals' favourite, a dish they relish between lips and teeth.

When it comes to seafood in the Liaodong region, Dalian stands out from the rest of the regions. Compared with the characteristic food in Jinzhou, Panjin and Dandong , seafood cuisine in Dalian is relatively more systematic and pioneering. That is why it is often referred to as Dalian Cuisine—a separate cuisine in its own right. As an original cuisine, Dalian Cuisine is the first cuisine in China that has seafood as the main raw material. Dalian Cuisine originated from Shangdong Cuisine and systemically absorbed some characteristics of Japanese and Russian Cuisine giving rise to its own and distinctive style. Though it is original, Dalian Cuisine still bears a historical origin—Venturing to Northeast China.

From the 19th century to the middle of the 20th century, a significant number of people from Shandong, Hebei, Anhui provinces, and other places migrated to the northeast of China. Among these people, most are from Shandong Province and they have exerted the biggest influence on Dalian. The large-scale population migration also brought changes and developments to the cooking culture, and Dalian Cuisine has gradually come into being in the course of history. During this period, some typical Dalian food emerged including *salted fish pie (Chinese style)*. To cook this dish, knead cornmeal into dough and bake it to a crisp Chinese pancake. Then cut it into a few pieces and set it aside. Then

谈起辽东的海鲜美食，大连是一个不可忽视的代表。与锦州、盘锦、丹东的特色食物相比，大连的海鲜美食更具有体系，也更能表现开拓精神，而它也有属于自己的名字——连菜。连菜是大连独创的菜系，是全国第一个以海产品为主要原料的菜系，其根源是鲁菜，也吸纳日餐和俄餐的特点，形成独具特色的菜系。连菜虽然是独创菜系，但也有着历史渊源。与连菜的形成紧密相关的历史是闯关东。

从十九世纪到二十世纪中叶，山东、河北、安徽等地的人口大量迁移至东北，其中以山东人数量最多，对大连影响最大。规模巨大的人口迁移也给饮食文化带来了改变，连菜在历史的进程中逐渐形成。在这一过程中，出现了一些颇具代表性的食物，例如咸鱼饼子。用玉米面和成面团，在大锅边贴成一面焦脆的大饼子，饼子熟后切成象眼块备用，再将咸鱼切成条炸熟备用。最后以葱姜烹锅，加入青红尖椒煸炒，加入咸鱼和饼子，加料酒、味精、少许糖炒透，加花生米炒匀，淋香油即成。饼子的面香与咸鱼条的咸香混合，别具风味。这道食物也在粮食与海鲜的碰撞中显露出饮食文化融合的一抹痕迹。

连菜能够在美食争奇斗艳的中华土地上逐渐崭露头角，离不开对鲁菜的传承，但更多的是依靠其独创的特色。新中

cut the salted fish into strips and fry it until it is done. Finally boil scallions and ginger in a pot, add green and red chilli and stir-fry. Then put the Chinese pancake and salted fish into the pot and add cooking wine, MSG, a little sugar, as well as peanuts and stir well. Finish by adding a touch of sesame oil. The noodle's smell is mixed with the salty flavor of fish bringing an unprecedented flavor to your tongue. As an integration of grain and seafood, this dish is also an embodiment of the integration of different diet culture.

Dalian Cuisine, relying more on its original features than on the inheritance of Shandong Cuisine, stands out among Chinese Cuisines, an amalgam of delicacies. Sea cucumber with fried chicken leg, Pacific abalone with scallop, agate-coloured shrimp and pearl-shaped fish, which are four original Dalian dishes, have definitely made their way onto the state banquet menu. With regard to the popularity of Dalian Cuisine, the Master Mou Chuanren deserves much of the credit. Yantai city of Shandong Province is the birthplace of Master Mou who has long been considered as the mainstay of Dalian Cuisine and an expert in cooking Shandong Cuisine. Based on the philosophy of sincerity, authenticity and delicacy, Master Mou himself, on behalf of Dalian, participated in the first-ever National Competition of Culinary Art of China. After a week of competition, he finally got the Gold Award for Dalian after outperforming numerous culinary masters from the Eight Great Cuisines of China. His success enabled Dalian Cuisine to swiftly obtain the fame it deserved.

These four dishes that have been served at the State banquet have their own characteristics. Sea cucumber with fried chicken leg

国成立以来，就有连菜中的独创菜品鸡锤海参、鲜贝原鲍、橘子大虾、红鲷戏珠被奉为国宴菜，登记入册。而这些菜肴能够入选国宴菜单，也有一段由来。身为山东省烟台人的牟传仁大师是鲁菜大家，也是传承连菜的柱石，他坚持着最正宗的传承、最精致的出品、最执着的理念，只身一人代表大连参加了新中国成立以来的第一届全国烹饪大赛。经过一周苦战，力拼全国八大菜系高手，最终为大连带回了金奖，为连菜打出了名头。

作为国宴菜，这四道菜各有特色，鸡锤海参注重的是色泽明亮，质地软滑，味醇鲜香，酥烂适口。橘子大虾突出的是红润油亮，色如玛瑙，食之鲜甜适口。红鲷戏珠则白、黄、绿三色相间，交相生辉，原汁原味，清淡鲜香。而鲜贝原鲍刀工讲究，制作精细，鲜珍海味营养丰富，盛装别致，独具一格。

鲜贝原鲍一菜，要用大连金州区出产的一种名贵海产贝类——皱纹盘鲍为主料。皱纹盘鲍是我国所产鲍中个体最大者。这道菜主料为新鲜的皱纹盘鲍、鲜扇贝肉。辅材有冬笋、香菇、青豆、葱、姜、蒜。调料有葱油、绍酒、味精、湿淀粉、清汤、精盐、椒油等。制作时首先将鲍鱼壳刷洗干净，用刀尖沿壳边将肉挖出，摘出脏腑，片去壳肌，空鲍鱼壳用沸水

focuses on the bright color, soft and smooth texture, and mellow and fresh taste. Agate-coloured shrimp is prominent for its sweet and fresh taste and reddish shiny color which looks just the same as agates. Pearl-shaped fish emphasizes both its alternating colors between white, yellow, and green and the relatively light and fresh mouthfeel without losing the original flavour. Based on the exquisite knife skills and highly sophisticated cooking methods, Pacific abalone with scallop is remarkable for its nutrient-rich seafood and unique style as well.

The main ingredient of the Pacific abalone with scallop is Pacific abalone, the biggest domestic abalone which is a precious seafood produced in Jinzhou District of Dalian. Other ingredients include fresh scallop meat, winter bamboo shoots, mushrooms, green beans, scallions, gingers and garlics. The seasonings include scallion oil, Shaoxing wine, MSG, wet starch, clear soup, refined salt, and pepper oil. To cook the dish, clean abalone shells, then use the tip of the knife to scoop out the meat along the edges of the shell. After that, remove giblets and wash clean, then arrange the empty abalone shells to wedge around the plate after boiling. Next, carve out the shape of chrysanthemum where the meat is thick, cut the half from the middle, dice half of the winter bamboo shoots and slice the rest to diamond shapes, then dice the scallions and ginger and slice the garlic. Thereafter, pour water and a litter of oil into a wok and heat it. Then add pacific abalone and scallop meat to boil to medium rare, then drain them. After that fry it in hot oil with a temperature of 150 degrees then mix chicken soup, Shaoxing wine, MSG, wet starch, salt until the sauce turns into white. Set it aside

煮后，摆在盘子边沿。取鲍鱼肉在贴壳面剞菊花刀，一切两半。冬笋、香菇一半改丁，一半切象眼片。葱、姜改末。蒜切片。其次将汤勺置火加清水烧沸，下鲍鱼、鲜贝，余五成熟捞出，沥干水分，再分别入六成热油中稍炸出锅。用鸡汤、绍酒、味精、湿淀粉、盐兑成白汁芡，以上述调料再加酱油成红汁芡待用。炒勺下葱油，用姜蒜炝锅，下冬笋、香菇丁、鲍鱼快速颠翻，烹入白汁，加椒油颠匀出勺，装鲍鱼壳内。最后炒勺刷净，置火下葱油，葱姜蒜炝锅，下青豆稍炒，速入鲜贝炒转，烹入红汁，加椒油颠匀出勺，盛入圆盘中心。

与辽东人嗜肥浓、喜腥膻、重油偏咸的传统口味相比，创新的连菜更加注重海鲜食材原本的味道，对于汁浓芡亮、鲜嫩酥烂、形佳色艳等特点则予以保留。大连地理优势得天独厚，加之对历史的传承和面向未来的开拓创新，孕育出连菜这一新秀。二〇〇八年，大连市政府组织一百多位厨艺大师和专家共同起草《连菜烹调操作技术规程》，对一百道大连菜品的主辅料配比、制作工艺和流程进行了规范。连菜有了正宗的做法，百年历练，终成正果。

在中国漫长的海岸线中，辽东所处位置独一无二。但在世界范围内，相似的地理环境却并不鲜见，且都孕育了当地的特色美食。跨越朝鲜半岛，日本的新潟有种叫冲汁的乡土

and then mix the soy sauce with above seasonings until the sauce turns red. Again, set it aside and then combine scallions, gingers, garlics and oil; pour over winter bamboo shoots, mushrooms and abalone to sauté quickly and add white sauce and pepper oil to stir well. Serve them in the abalone shell. Finally wash the wok, and combine scallions, ginges, garlics and oil. Add green beans and diamond-shaped winter bamboo shoots to saute a little and quickly put scallops into the wok and stir-fry. Then add red sauce and pepper oil to stir well and arrange them in the center of the plate.

Compared with the traditional taste of Liaodong people who prefer fatty and fishy food with high intake of salt and oil, the newly created Dalian dishes offer more focus on the original flavour of seafood. However, some quintessential features such as attractive taste, colourful and seductive appear-ance are preserved and reflected in the current Dalian Cuisine. As a peninsular city, Dalian connects with the northeast mainland of China. It has a unique geographical advantage, rich historical inheritance, and future-oriented innovation that make Dalian a thriving city—the city where Dalian Cuisine was born. In 2008, Dalian municipal government brought together over 100 master chefs and experts to draft the Standard Cooking Skills and Procedures, which standardized the proportion of main and auxiliary ingredients, cooking techniques and procedures of 100 Dalian dishes. Since then, Dalian Cuisine has had its orthodox cooking methods through hundred years of development.

Liaodong, like many other places around the world with the similar geographical location, seems to be the cradle of local

料理，当地所产的芋头混以其他鲜蔬与鲜鱼同煮，在不同人家的厨房，烹出不同的味道，饶有趣味。而在美国东海岸缅因州，当地所盛产的波士顿龙虾，则在厨师的手中以不同的菜式呈现出来，享誉世界。在大西洋东岸的拉科鲁尼亚，则可以品尝到有别于经典西班牙海鲜饭的拉科鲁尼亚海鲜饭，当地的海鲜饭味道偏甜鲜，于奶酪香气中突出海鲜的海味。相似的地理环境，不同的文化，不同的习惯，造就了各具特色的海鲜美食。

目光再回到辽东半岛，鲍鱼、刺参还蛰伏在深水之中，静静生长。带着希望的渔船已经扬帆出海。劲爽的海风裹挟着大海的味道扑面而来，辽东的人们在这熟悉的味道中，将继续迈着传承与开拓的脚步前进。

specialties. Across the Korean Peninsula lies Niigata, a Japanese city in which a local dish named Chongzhi was born. Local taro cooked with fresh vegetables and fish differs in taste from home to home. While in Maine, a state in the northeastern of United States, delicacies made from locally produced Boston lobsters are well known around the world. As for La Coruna on the east coast of the Atlantic Ocean, the paella has spread by word-of-mouth because of its harmonious blend of cheese and seafood. Similar geographical conditions, together with different cultures and habits, have given birth to seafood delicacies which all have various distinctive features.

On Liaodong Peninsula, where abalone and sea cucumber quietly grow in the deep waters. The fishing boats with hope have set sail. Then spanking sea breeze comes along with the familiar sea flavor, through which Liaodong people continue to forge forward with the pace of inheritance and development.

传统，津门的奢华

第二章

Chapter

Two

*Tianjin Cuisine:
Traditional Luxury*

　　《晋书·天文志》载：天津九星，横河中，一曰天汉，一曰天江。所言之意为星官天津是银河的渡口。自古以来"津"就有着渡口的含义。明建文二年，朱棣将天子渡过京杭大运河之处赐名为天津，自此就有了这座城市沿用至今的名字。天津汇河入海，又是京畿要地，在经济与军事发展的同时，也书写着独特的海洋美食故事。

　　从地理上讲，天津东临渤海，西扼九河，因此水产丰富，且品质优良。从历史沿革来讲，天津商周时已有人居住，以捕鱼为生；汉武帝时设置盐官，天津盐业发展起来；隋时，京杭大运河建成，天津以三会海口闻名于世，漕运兴起；明时，始设天津三卫，其水陆要冲的地位凸显；晚清至近现代，天津屡遭磨难，但商贸繁荣发展。从历史中走来的天津，在传统中彰显着它的品味，而丰饶与品位的碰撞，则是津门传统中的奢华。

The *Book of the Jin Dynasties: Astronomy* states that there are nine stars in the charge of Tianjin Xingguan.[1] They divide the Milky Way into two, one called Tianhan and the other Tianjiang. This division indicates that Tianjin Xingguan is the ferry of the Milky Way, and this is also why the Chinese word 津 (*Jin*) has, for a very long time, had the meaning of ferry. In AD1400, Zhu Di, the then emperor's uncle who became the next emperor, led his troops cross the famous Beijing-Hangzhou Grand Canal, and he decided to call the canal where they crossed Tianjin. Ever since, Tianjin has become the name of this northern city. As a city which is extremely close to the capital Beijing, and where the nine local rivers meet with the Bohai Sea, Tianjin has also written a unique chapter of seafood stories.

Geographically speaking, Tianjin lies to the west of the Bohai Sea and to the east of nine local rivers. This has significantly contributed to the rich variety and the excellent quality of its aquatic products. Historically speaking, people start to live here and make a living by fishing as early as the Shang and Zhou Dynasties. Emperor Wu of the West Han Dynasty gave orders to set a position named Salt Official in charge of salt affairs, and since then, the salt industry in Tianjin began to develop. During the Sui Dynasty, the Beijing-Hangzhou Grand Canal had been completed thereby facilitating the prosperity of water transportation. In the

[1]

Xingguan (星官) is a product of the combination of ancient Chinese mythology and astronomy. In order to facilitate the recognition and observation of stars, ancient Chinese astronomers grouped several stars, each named after a thing on the Earth. The group is called Xingguan. But this constellation does not contain the meaning of the starry zoning, which is different from the concept of constellation as we know today.

菜品的传统意味着传承，奢华则包含用料、品相、滋味、技法等诸多内容。天津菜在这些方面都颇具特色。天津人居史悠久，传承自不必说。明清两朝六百年，天津拱卫皇都，奢华也有由来。清朝，天津更被称为食都，津菜更是以"宫、商、馆、门、家"著称，也就是我们所说的宫廷菜、商埠菜、公馆菜、宅门菜和家庭菜。

我们回溯历史，从宫廷菜说起。明正德年间，太监刘瑾曾在天津设"银鱼厂太监"，专门搜罗海鲜与河鲜，送到京城，供皇帝及宫廷食用。天津银鱼之珍贵由此可见一斑。银鱼淡水海水中皆有出产，但金睛银鱼为天津所独有。其鱼体呈圆柱形，无鳞，全身透明如玉，通体一条软骨，刺少而细软，熟后鲜嫩异常。今天，金睛银鱼几乎绝迹，但以渤海出产的普通银鱼做出的高丽银鱼仍是一道天津的传统名菜。

将银鱼掐去眼睛、尾梢，洗净，沥去水分，用精盐腌制片刻后滚匀面粉备用。将鸡蛋清迅速抽打成泡沫状，随后分三次撒入淀粉搅拌均匀，制成"高丽糊"。中火热油至五成熟，离火，将银鱼逐条放入蛋泡糊中拖匀，下入锅中。然后，移锅用小火将银鱼慢炸至浅黄色，捞出控油，入盘。最后汤锅置旺火上，下入清汤、食盐、味精烧沸，以湿淀粉勾芡成白汁，舀入小碟内，与花椒、盐、辣酱油、银鱼同上桌，供

Ming Dynasty, Tianjin Wei● was established, bringing to prominence its water and land transportation. From the late Qing Dynasty to the modern era, Tianjin has suffered from many hardships, but trade has flourished. Tianjin, which has a long history, continues to highlight its taste for tradition. The collision of richness and taste has given rise to the luxury of Tianjin Cuisine.

The tradition of Tianjin Cuisine refers to its continuation, while its luxury refers to its ingredients, appearance, taste, cooking skills, etc., all of which have made Tianjin Cuisine special. Tianjin has a long history, so it is natural that it has inherited many traditions. In the six hundred years of the Ming and Qing Dynasties, Tianjin surrounded and protected the capital city, leading to the luxury of its cuisine. In addition, Tianjin was called *Food City* during the Qing Dynasty, and Tianjin Cuisine was also famous for its imperial dishes, commercial dishes, mansion dishes, rich family dishes and common home dishes.

At this juncture, it may be relevant to look back beginning with the imperial dish. Between AD1506 and AD1521, the eunuch Liujin, set up the Whitebait Department in Tianjin to collect seafood and river fishes for the emperor and royal family members. This endeavor enables us to perceive the preciousness of the Tianjin whitebait. Whitebait breeds both in freshwater and saltwater, but golden-eyed whitebait is a specialty of Tianjin. This kind of fish is cylinder-shaped, devoid of scales and transparent. The fish has fine and delicate flesh and it also has one soft bone in the whole body and very few small bones. Unfortunately, the golden-eyed whitebait is almost extinct nowadays. That notwithstanding, the normal

●

Wei(卫) is the military system of the Ming Dynasty. Each Wei has a total of 5,600 soldiers, and the Tianjin Wei includes three Wei with a fixed number of 16,800 soldiers.

择而蘸食。今天，人工饲养繁殖增多，银鱼已走上普通人餐桌，大家在分享银鱼美味之时或许可以体会到当年宫廷菜的那份奢华。

天津渔业、盐业、漕运皆兴，商贸发达，其商埠菜自然是名声斐然。在"宫、商、馆、门、家"中，商埠菜最为丰富，最能体现天津菜的特点，也最能看出历史的变迁。在天津，商埠菜的代表是八大成饭馆，即聚庆成、聚和成、聚乐成、义和成、义升成、福聚成、聚升成、聚源成。其中聚庆成历史最久，于康熙元年开业。此后两百余年间，天津有名的菜馆增增减减，但商埠菜并未消失，商埠菜的厨师们，擅烹两鲜、讲究时令、精于调味、技法独特，使得天津菜的特点完美展现。擅烹两鲜，在酸沙类菜品中表现突出，如名贵的酸沙紫、酸沙目鱼片；讲究时令，则春吃黄花鱼，夏吃鳎目鱼，秋吃海蟹，冬吃银鱼，间而有各种虾类、贝类、海珍，以至于民间有"一鲙，二平，三鳎目"之说，菜品有鲍鱼虾、煎烹大虾等。关于技法、调味、口感，则可通过扒通天鱼翅一览真容。

扒通天鱼翅是"八大成"的一道特色名菜。"扒"本不是津菜所独有的烹饪技法，但天津人在时间中创造了自己的特点，有了勺扒这一技法。操作时需运用大翻勺的技艺，讲

whitebait produced in Bohai Sea is the main raw material in the traditional Tianjin dish known as Koryo Whitebait.

To prepare the dish, it begans by removing the eyes and tails of the whitebait and washing the fish clean. Next, drains the water, marinates the fish with salt, and rolls them in dry flour. Second, stir egg white into foam, sprinkle three rounds of starch on it and mix thoroughly to produce Koryo paste[1] or egg foam paste. Then heat oil in the wok, put whitebait in the Koryo paste and then into the wok one by one. Fry the fish until they turn golden, then take them out, and put them into a plate. Third, put a soup pot on the fire, add some clear soup, salt, MSG, and when it starts boiling, add some wet starch then stir. Ladle out some soup as saucers, and serve the saucers, the fish, the pungent sauce, the pepper and the salt together for people to enjoy. Today, thanks to the artificial cultivation, the whitebait fish is no longer a rare thing for ordinary families. This makes it possible for people to experience the luxury of imperial dish while enjoying the current delicacy.

Given the prosperity of the fishing industry, salt industry, and water transportation, Tianjin has a very developed trade and commerce sector that has built up the reputation of its commercial dishes. Among the five dish types, commercial dishes are the one richest in variety, and they best demonstrate the features of Tianjin Cuisine and the historical changes that have taken place in the city.

In Tianjin, the representatives of commercial dishes are the Eight Great Cheng restaurants, namely Juqingcheng, Juhecheng,

[1]

Koryo(高丽) is an ancient country located in Korean Peninsula (AD918–1392). Its territory roughly covered the middle and south of today's Korean Peninsula. Young girls in Koryo always wore beautiful white dresses. As people thought the color of egg foam was similar to that of the dresser, they started to call egg foam paste "Koryo paste."

究上下翻飞、左右开弓，大至整鸡、整鸭、三四条熬熟的鱼，小到细如火柴梗的白肉丝，都要能翻动自如，因此又称之为"勺扒"。扒通天鱼翅的主料是鱼翅，取整只排翅，入锅煮软，拆去翅骨，翅肉加猪肘、鸡翅膀、碎肉骨、葱、姜、绍酒、酱油、白糖、味精，上笼蒸五小时左右取出。随后，鱼翅入高汤稍余控干，投入用葱段炸香的锅内，加料酒、酱油、高汤、白糖，烧沸后移微火煨至汤汁稠浓。另用炒锅，下猪油少许，将剩余卤汁、调料倒入，再放鱼翅。烧沸后，转用小火略烤，下湿淀粉勾芡，滴上葱油，大翻勺，淋上明油装盘即成。因取用全只整翅烹制，又兼技法独特，品相口感皆为上等，为津菜的"百菜之王"，因此取名"通天鱼翅"。只扒通天鱼翅一道菜，津菜的奢华便不言而喻。

在中国近代史中，天津有着极为重要的地位，从晚清至北洋政府，再到民国，天津聚集了各路军阀、政客、富商，因此公馆林立。虽然是在动乱的时代，公馆中的饮食仍旧有一番讲究，天津也就有了公馆菜。其中荣益海就是一位从公馆中走出的厨师。荣益海是河北省三河县人，学徒有了名气之后，曾在马步芳公馆和曹锟公馆做过家厨。虾籽狮子头就是出自他手的一道菜。这道菜需以大沽口应时鲜虾为主料，将肥瘦相间的上好猪肉、火腿、虾籽、荸荠、笋丁拌匀后挤

Julecheng, Yihecheng, Yishengcheng, Fujucheng, Jushengcheng, and Juyuancheng, of which Juqingcheng is the oldest because it was opened in AD1662 during the Qing Dynasty. In the following two centuries and more, well-known restaurants in Tianjin kept coming and going, but commercial dishes have not disappeared. The chefs and cooks are good at seasoning, cooking both seafood, river fishes, and preparing various dishes according to different seasons with unique cooking skills. All of these contribute to perfectly exhibiting the characteristics of Tianjin Cuisine. A series of dishes whose names include the word Suansha[1] are always cooked with both seafood and river fishes. The dishes include the luxurious suansha purple crab and suansha flounder slices. Due to their sensitivity to the seasons, Tianjin people usually eat yellow croaker in spring, sole in summer, sea crab in autumn and whitebait in winter. In addition, they have various kinds of shrimp, shellfish, and other seafood. Typical dishes include abalone and shrimps and fried prawns. As for the seasoning and cooking skills, the tip of the iceberg is perceptible through the famous braised shark's fin dish.

Braised shark's fin is a famous dish of the Eight Great Cheng restaurants. As a matter of fact, "braise" is not a unique cooking skill manifested by Tianjin Cuisine, but Tianjin people have created their own characteristics, and with time, the scoop-braising skill has stood out. The skill requires proficiency in using a large spoon to flip (up and down) and sway (back and forth) materials as large as a whole chicken, a whole duck, three or four cooked fish, or as

[1]

Suansha (酸沙) is a special cooking skill in Tianjin. This kind of dish pays attention to sour, sweet, slightly salty and slightly spicy flavor, and it has the taste of mashed potatoes.

small as white pork shreds that are as fine as matchsticks. The main ingredient is the fin of a shark. To cook this dish, put a whole fin into a wok and add some water to boil till the fin becomes soft. Remove the bones of the fin, and add pork knuckles, chicken wings, minced pork and bones, scallions, ginger, cooking wine, soy sauce, sugar and MSG to the fin. Steam all ingredients for about five minutes. Later, simmer the fin for a while in a broth pot then dry it. Then, put scallions, the dried fin, cooking wine, soy sauce, broth, and sugar into the heated wok with oil. After boiling, turn the fire down and simmer until the soup gets thick. Finally, put a little lard in another wok and heat it. Pour the remaining soup and seasoning into it, and then put the fin in. After the wok starts boiling, reduce the fire to stew for a while, then add some wet starch and stir. Finish by adding a touch of scallion oil. Because it is made from a whole fin and by special skills, the dish, with its superb taste and appearance, is praised as the King of Tianjin dishes. The luxury of Tianjin Cuisine is self-evident through this dish.

In the modern history of China, Tianjin has an extremely strategic position. From the late of the Qing Dynasty through the government of the Northern Warlords of China to the Republic of China, Tianjin had witnessed various warlords, statesmen, and wealthy businessmen boosting the mansion dishes. Despite the era of turmoil, people living in mansions still enjoyed high-quality food. Rong Yihai is one of the chefs who used to work in mansions. Born in Sanhe County, Hebei Province, Rong Yihai worked as a chef in Ma Bufang Mansion and Cao Kun Mansion after he became famous. He was particularly skilled in preparing

成球形，滚馒头丁炸成，成才色泽金黄、外酥里嫩。狮子头原本属于淮扬菜，在天津鲜虾和虾籽的加入下，虾籽狮子头则显现出天津的特色。

天津的宅门菜、家庭菜虽不像宫廷菜、商埠菜、公馆菜那样讲究奢华，但也足够丰富，而这种丰富对于寻常人来说就是平凡生活中的奢华。旧时普通的天津人不常去八大成这样的地方，但二荤馆、酒席处却是可以去的，其代表是八大碗、四大扒，这其中用到海鲜的就有不少，如细八大碗中的烩虾仁，粗八大碗中的蛋羹蟹黄、海参丸子，四大扒中的扒海参等。但更能代表宅门菜或者家庭菜的则是一道"津门全爆"。

津门全爆以多种原料集于一肴，并运用爆的烹调技法制成。主要原料是天津常见的各类海鲜，包括虾仁、全贝、海参、鱿鱼、鱼肉等，并辅以鸡肉、肚头、香菇、玉兰片、荸荠等。材料洗净，荤类切丁，以精盐、蛋清、淀粉腌制。素类切小象眼片并用开水焯过后控干。另取一个碗放入清汤少许，加适量湿淀粉、绍酒、味精及精盐调成汁。炒锅内放熟猪油，中火烧至六成热，先将虾仁、全贝、海参、鱿鱼、鱼肉、鸡肉倒入，再下肚头，稍拨动即捞出。炒锅内留底油稍许，旺火烧热，放葱、姜、蒜末炸出香味，下香菇、玉兰片、

fried minced shrimp ball which became one of his representative dishes. To make the dish, first, chop fresh shrimps from the local market into paste, then mix it with chopped pork, ham, shrimp roe, diced water chestnuts and bamboo shoots, and squeeze the mix into balls. Finally, roll the balls in diced steamed buns before frying. The balls are ready when they turn golden. Taste one of them to ensure that first, it tastes crisp outside and then tender inside and it smells good. Originally, the minced pork ball dish belongs to Huaiyang Cuisine❶ , but with the addition of Tianjin shrimps and shrimp roe, the fried minced shrimp ball perfectly displays features of Tianjin Cuisine.

Rich family dishes and common home dishes are not as luxurious as imperial dishes, commercial dishes and mansion dishes, but they also have a rich variety, and such richness constitutes the luxury of ordinary life for ordinary people. In the old days, ordinary Tianjin people did not have plenty of opportunities to go to such places as the Eight Great Cheng restaurants and they could only afford to eat in smaller restaurants. They typically ate Eight Large Bowls and Four Braises dishes that were predominantly seafood dishes including stewed shelled shrimp, custard and crab roe, minced sea cucumber balls, braised sea cucumbers. But there is one more representative dish known as Tianjin stir-frying.

The main materials are a large variety of common seafood in Tianjin which includes shelled shrimps, scallops, sea cucumbers, squids and fish, and they are supplemented by chicken, pork tripe, mushrooms, bamboo shoots, and water chestnuts. First, rinse all the raw materials, dice all animal meat, and marinate the meat

❶

Huaiyang Cuisine（淮扬菜）is one representative of Jiangsu Cuisine which we are going to talk about in one of the following chapters.

荸荠等，倒入汁芡，颠翻出锅即成。食用时外带小碟虾油上桌。津门全爆讲究出菜干净，颜色鲜亮。其成品色艳，鲜香脆嫩，回味无穷。

津菜之中，清真菜也很有名。于清咸丰年间开业的天津鸿宾楼，是一家久负盛名的清真风味饭庄。清真菜并不常用两鲜，但鸿宾楼却有一道"火笃鱼白"的精美菜肴，宾客为此纷至沓来。此菜是用渤海所产河鲀、黄鱼或鳕鱼的鱼白所做，鱼白难求，无法应市。"火笃鱼白"又打动了人们的味蕾，因此鸿宾楼名厨们屡经试制终于想出新招：将刀鱼剥皮去骨剁成鱼泥，制成牛舌状，代而烹之，名曰"独鱼腐"。结果，不仅其外形极似鱼白，而且色泽以假乱真。因其质地柔嫩犹如豆腐，入口即化，遂定名为"鱼腐"。这也称得上是一段趣闻，津菜特有的技法"笃"也因此为人熟知。

天津在晚清至抗战胜利的这段历史时期内，政治地位十分重要，市区内前后曾有九个国家的租界。租界的存在，给天津带来了磨难，也无意中给天津带来了国外的文化，这些文化的影响在饮食上也有所体现。1863 年，北方最早开业的西餐馆就是天津利顺德大饭店。原本做西餐的都是外国厨师，但到二十世纪初，天津厨师已可以独自完成西餐大菜，并融合了天津菜的技法和口味特点，比如利顺德的双冬烧海

for a while by adding salt, egg white, and starch. Meanwhile, cut all other materials into diamond shape and then blanch. After that, pour a little broth into a different bowl and add wet starch, Shaoxing wine, MSG, and salt to make sauce. Next, heat some lard in a hot wok at a temperature of 150 degrees with medium heat. Pour in shelled shrimps, scallops, sea cucumbers, squids, fish, chicken, pork tripe and remove after stir-frying. Keep a little oil in the wok, then put scallions, gingers, and garlics to fry. Add mushrooms, bamboo shoots, water chestnuts, etc. Finally, pour the sauce into the wok, stir well and arrange them in the center of the plate. Do not forget to serve the dish with shrimp oil. Tianjin stir-frying has a bright color and a clean appearance. It is fresh and crisp and it offers you an endless aftertaste.

Chinese Halal food is also well-known in Tianjin. Hongbinlou Reataurant, which opened in Tianjin in 1863 during the Qing Dynasty, is a long-standing Halal-style restaurant. Usually, seafood and river fishes were not common in Halal dishes, but the restaurant had a fine dish called stewed milts, which attracted lots of customers. However, this dish was cooked by using the milts of puffers, yellow croakers, or codfish obtained from the Bohai Sea, so it was not long before the supply of milts was unable to satisfy people's demands. Since the dish had already left an indelible taste in people's mouths, chefs had to create a new dish named stewed fish tofu after plenty of trials. To cook this dish, first, peel a knifefish, remove its bones and chop it into paste. Then make the paste into an-ox-tongue shape and cook it, making the shape and color both similar to milts. Because its texture was as soft as

参，就是用海参配以冬菇、冬笋、虾米焖烧而成，口感糯软，汁浓味鲜。在旧天津，会以渤海所产鲜鱼切薄片，做成刺身，往往会配以独特的蘸食酱汁。这实际上是天津菜对外国菜品的接受和融合。

今天，这座东临渤海、西扼九河的城市已经是世界等级最高的人工深水港，它无愧于"天津"这个有着悠久历史传统的名字，而在经济繁荣发展的未来，它也将保持人们舌尖的那份奢华。

tofu, people called it "Fish tofu. " With this anecdote, the unique stew(笃) technique of Tianjin Cuisine has since then become well known.

In the period from the late Qing Dynasty to the victory of the War of Resistance Against Japanese Aggression, the political status of Tianjin was very important. There were concessions of nine countries successively established in Tianjin and these concessions brought hardships to Tianjin while inadvertently introducing foreign cultures to this city. This cultural influence eventually reflected in people's diet. The Astor Hotel in Tianjin was the first western restaurant in north China after the Second Opium War (AD1856–1860). Originally, western food was cooked solely by foreign chefs, but by the beginning of the 20th century, Tianjin chefs had been able to prepare western dishes on their own and also combine the techniques of Tianjin dishes. For instance, braised sea cucumbers with mushrooms and winter bamboo shoots, the representative dish of Astor Hotel, is cooked with sea cucumbers, mushrooms, winter bamboo shoots and dried shrimps. It has a soft and delicious taste. Besides, fresh fish from the Bohai Sea was often sliced and made into sashimi, which in Tianjin is often accompanied with unique sauce. These modifications reflect how Tianjin Cuisine absorbed and improved upon western dishes.

Today, Tianjin, the city that embraces the Bohai Sea and gathers nine rivers, has become a world-class artificial deep-water port. It is worthy of its name given its long history and fine tradition. In addition, it has the potential to maintain the luxury on people's tongues in the prosperous future.

绵延，齐鲁的品格

Chapter

Three

Shandong Cuisine:
Continuing Values

　　周初分封天下，鲁居泰山之阳，齐居泰山之阴，至战国末，始称齐鲁。在历史的变迁中，齐鲁所指地域多有变化，清初才设山东省并沿用至今，而今天，齐鲁所指的就是山东。齐鲁大地地貌多元，物产丰富，这块土地上诞生的鲁菜则是一座美食的宝库。鲁菜是八大菜系中唯一的自发型菜系，被称为自发型菜系意味着鲁菜经历了独立而完整的起源、发展与传播，这就是鲁菜的绵延。那么，让我们走近一些，去发现鲁菜历史绵延中与海洋相关的痕迹。

　　春秋时，齐国国相管仲主张治国以民食为重，这成为鲁菜发展的契机。彼时，齐鲁之地名厨辈出，其烹饪技法已相对成熟。思想方面，孔子则提出了中华饮食的核心观念。所有因素结合在一起，成为鲁菜起源的基础。发展至秦汉，已有名膳列于食谱，如西汉时的鳀鲑。相传汉武帝攻打夷民的过程中，有次到达海边，忽然闻到一种特殊的香味。

ifferent kingdoms were shaped in the early Zhou Dynasty as a result of enfeoffment. When regional states became increasingly powerful, present day Shandong Province acquired the nickname Qilu, a name derived from the states of Lu and Qi, located in the south and the north of Mount Tai respectively. During the course of history, Qilu kept changing its jurisdictional areas until Shandong Province was established in the early Qing Dynasty. Shandong Province is richly endowed with resources and it boasts various landforms where Shandong Cuisine (Lu Cuisine) was born. As one of the Eight Major Cuisines of China, Shandong Cuisine appears to be the only self-created cuisine. It is rooted to the native cooking styles of Shandong, and has experienced a long process of development and spreading. That is what we refer to as "continuity" of Shandong Cuisine. So, it is only natural to go deeper to discover seafood related stories of Shandong Cuisine.

During the Spring and Autumn period, Guan Zhong, the prime minister of the State of Qi claimed that the top priority of governing a state should be placed on satisfying people's dietary needs. This provided Shandong Cuisine the necessary political base on which it developed. Since then, talented chefs with relatively mature cooking techniques have jostled in Qilu area. From a cultural perspective, Shandong Cuisine is infused with Confucian values of moderation, harmony and balance. On the basis of all these factors, Shandong Cuisine developed so quickly that famous dishes including Zhuyi (marinated fish intestines) made their way into the recipe during the Qin and Han Dynasties. According to the

后发现是渔民在炕中烹制鱼肠，其上覆土，成熟时香气外溢。武帝遂命人取来品尝，只觉味道鲜美，遂命名�。这道菜后来也成为汉代宫廷菜肴。两晋南北朝时期，贾思勰的《齐民要术》记载了许多食材加工的技艺。贾思勰是今山东寿光人，又曾任高阳郡太守，他的著述，从侧面说明鲁菜逐步发展起来。

有唐一代，饮食极为丰富，朝中甚至设有"鱼师"一职，掌管捕鱼，以供皇室享用。在典籍中则记载了海鱼类、贝类、虾蟹类、章鱼、海蜇类等数十种海鲜的烹调与食用方法。彼时，齐鲁大地归属于河南道，莱州、登州等地的鱼市成为鲁菜的取材地。至宋朝，民间的富庶超过盛唐，因此鲁菜也得到长足发展，如至今仍保留的一道鲁菜名膳扒鲍鱼在宋朝时已名声响亮。

元明两代，鲁菜中关于海鲜的菜肴在杂剧、小说等文学作品中有所体现，如《金瓶梅》第十一回，提到西门庆要吃银丝鲊汤，而这银丝鲊汤正是用渤海所产水母所制。清朝时，关于鲁菜中海鲜菜肴的记载则更多。郝懿行《记海错》一书载登莱海中产加吉鱼、海参，渤海海中产对虾。袁枚《随园食单》又道：须检小刺参，先泡去泥沙，用肉汤滚泡三次，然后用鸡、肉两汁红煨极烂。这种先水发，再用肉汤加热的

legend, Emperor Wu (141BC–87BC) of the West Han Dynasty was suddenly greeted by the enticing smell while he was on his way to chase Huns at the seaside where some fishermen were steaming fish intestines. They covered the fish intestines with soil and baked until they gave off a pleasant smell. The savory smell and the fantastic taste charmed Emperor Wu of the Han Dynasty who then named the dish Zhuyi and propagated it as a court dish. In the period of the Jin and Southern-Northern Dynasties, some cooking techniques were described in the book *Arts for the People*[1] written by Jia Sixie (a native of present day Shandong) through which we are able to witness the development of Shandong Cuisine in a different way.

In the Tang Dynasties, seafood dishes were so popular that the imperial court created a position for fishers whose job was to exclusively serve the royalties. As a result, several ancient books were written about the cooking and the ways of eating dozens of seafood such as sea-fish, shellfish, shrimp and crab, octopus and jellyfish. At that time, when Qilu area was under the jurisdiction of Henan supervisory region (including present day Shandong and Henan Provinces), the fish markets of Lai Chow (now Laizhou city in Shandong Province) and Teng Chow (now Yantai city in Shandong Province) became major sources of seafood. In the prosperous and wealthy Song Dynasty, Shandong Cuisine developed so rapidly that a large number of famous dishes sprung

[1]

Arts for the People was written by Jia Sixie in the Northern Wei Dynasty (AD386–534). It is the earliest outstanding agronomy and food science work completely preserved in China and the world. It records then agricultural production and food manufacturing in the Yellow River Basin. The original book, with extensive and rich content, consists of ten volumes and ninety-two articles, covering farming, forestry, animal husbandry, fishery, brewing and processing, as well as special description of cooking techniques.

发制方式，正是齐鲁之地所流行的。

从西周至清朝，上下三千年，鲁菜绵延不绝，随着时间的流淌而愈发具有生命力，鲁菜中与海鲜相关的丝丝缕缕也因此显得格外富有趣味。鲁菜的绵延不绝，因其有着最为重要的灵魂，这一灵魂就是鲁菜的文化与哲学，是鲁菜的品格。

齐鲁之地是儒学的发源地，鲁菜中的文化与哲学正源于儒学。《礼记》载：身修而后家齐，家齐而后国治，国治而后天下平。在修身齐家治国平天下的渐进过程中，须以"礼"为纲，这是儒学的核心思想之一，也是鲁菜文化哲学的要义。鲁菜主要流派有济南菜、胶东菜和孔府菜，而"礼"在孔府菜中尤为突出。

以宴喻礼是孔府菜的特色，如鱼翅席、海参席等。依古礼，设宴席处应隆重，依礼入厅，落座，随后看席。看席过后，净面，用到奉，继而茗叙，茗叙结束，方才入席。入席后，先用鲜果，之后随冷荤、热荤喝酒，酒过三巡，上正席。正席以大件配行件，鱼翅席第一大件上鱼翅，接着跟两个炒菜行件；第二个大件上鸭子，跟两个海味行件；第三个大件上鲑鱼，跟两个淡菜行件；第四个大件上甘甜大件，如苹果罐子，后跟两个行菜。少顷，上一甜一咸两

up. Till this date, some of these dishes (braised abalone with mushroom, for example) have remained star dishes.

As for the Yuan and Ming dynasties, seafood dishes appeared regularly in literary works including Zaju (a kind of Chinese drama) and fiction. For example, in the eleventh chapter of *The Golden Lotus*[1] , Ximen Qing (the hero of the fiction) said that he wanted shredded pickle fish soup — a dish made with jellyfish obtained in the Bohai Sea. During the Qing Dynasty, there were more depictions of seafood dishes in Shandong Cuisine. *A Record of Seafood*, written by Hao Yixing stated that the Denglai Sea was rich in red snappers and sea cucumbers while the Bohai Sea produced prawns. *Recipes in Sui Garden*, written by Yuan Mei claimed: we need to select little sea cucumbers and wash them clean. Soak them three times in hot broth and then braise them mixed with chicken broth and gravy until they are soft enough. The cooking method that consists of soaking food in water and then heating it with broth was widely used in the Qilu area.

From the Western Zhou Dynasty to the Qing Dynasty, the 3,000-year-long inheritance endowed Shandong Cuisine with growing vitality. Seafood in Shandong Cuisine is, therefore, of particular interest and taste. The continuation of Shandong Cuisine relies on its culture and philosophy, or in other words, on its style.

Qilu area gave birth to Confucianism which has a strong cultural and philosophical influence on Shandong Cuisine.

[1]

Jin Ping Mei (金瓶梅)—translated into English as *The Plum in the Golden Vase* or *The Golden Lotus*—is a Chinese novel of manners composed in vernacular Chinese during the late Ming Dynasty (AD 1368–1644). The author took the pseudonym Lanling Xiaoxiao Sheng (兰陵笑笑生), "The Scoffing Scholar of Lanling," and his identity is otherwise unknown, the only clue being that he hailed from Lanling County in present-day Shandong.

盘点心。接着再上四个瓷鼓子，再后上四个素菜，紧跟四碟小菜，最后上面食。食毕，奉牙签、槟榔、蔻仁，供客选用。最后再上洗脸水。整个宴席，礼法多而不乱，以食序礼，以食喻礼。

具体到菜肴，在孔府菜中有名的鱼翅菜肴数量颇丰，一如丁香鱼翅。乾隆年间，皇帝驾抵曲阜祭祀，因油腻而胃口不佳。孔府厨师无奈之中，随手将豆芽佐花椒爆炒，送上宴席，皇帝却被这道清爽小菜吸引，并连连称赞。自此炒豆芽便成了孔府的传统菜。后来在豆芽中加进了鱼翅。因豆芽掐去芽根，剩下的部分恰似白色的丁香花，配以鱼翅，犹如根根针穿小花，白中闪亮，新颖素雅，故有"丁香鱼翅"之称。

儒家维护"礼"，提倡"德"，重视"仁"。在践行这些儒家要义时，需要的是明辨是非，"中庸"也就成为应有之意，体现在鲁菜当中，就是五味的调和。对于厨师来说，五味的调和，是大智慧，需要大功力。鲁菜不同流派的厨师，各有法门，济南菜的厨师就常在汤菜中显示五味调和的技艺。如奶汤鲍鱼羹一菜，主料为鲍鱼，辅料为猪蹄、滑子菇、母鸡、香菜、菊花、猪瘦肉、猪胫骨、豌豆。制作时，先将猪蹄劈开。母鸡、猪精肉、猪胫骨洗净，用开水焯过，放大汤桶内

According to *The Book of Rites(Li Ji)*, the ancients who wished to demonstrate their illustrious virtue throughout the kingdom would govern well their states first. Wishing to govern well their states, they would regulate their families first. Wishing to regulate their families, they would first cultivate themselves. Confucius described "*Li*" (Rites) as a standard of conduct, which became the cultural essence of Shandong Cuisine. Shandong Cuisine is metaphorically referred to as a towering tree that consists of three main schools—Ji'nan Cuisine, Jiaodong Cuisine and Kongfu Cuisine, among which Kongfu Cuisine was the most luxuriant foliage. And Kongfu Cuisine attaches significant importance to "*Li*"(礼) .

Characterized by "manifesting '*Li*' through banquets", Kongfu Cuisine includes "shark fin banquet," "trepang banquet" and so on. The banquet is supposed to be in a glittering hall where guests enter in order according to ancient rites. No sooner do the guests sit than the dazzling exquisite plates full of auspicious snacks and fruit are displayed on the banquet table. Assistants then usher them to their seats after serving them face-washing water, appetizer and tea in turn. It is a characteristic of Kongfu banquet to be well-regulated and well-organized and this is specially displayed in the order in which dishes are served. Fresh fruit serves as appetizers, followed by cold dishes and hot dishes as well as liquor in that order. It is after several cups of alcohol that the main course, together with secondary courses, is served. For example, in the "shark fin banquet," the first main course is shark fins, followed by two stir-fried dishes; the second one is duck, with two seafood dishes followed; the third main course, in general, is salmon, after

which are two low-salt dishes; the fourth one however, must be a sweet dish like canned apple which is followed by two secondary dishes. After a while, two kinds of desserts are served — one is sweet while the other salty. After the dessert, four hot delicacies, four vegetables dishes, four plates of side dishes, in a specific order, are served before cooked wheaten food is finally served. Toothpick, areca nut, nutmeg and face-washing water are available for all the guests at the end of the meal. Complicated as its formalities are, the whole banquet is meticulously organized in a way that showcases the etiquette system and culture of Shandong Cuisine.

The numerous dishes of the famous "shark fin banquet" in Kongfu Cuisine are far beyond comparison. Take the dish white lilac shaped shark fin as an example. It is said that during the reign of Emperor Qianlong (AD1736 – 1795), at the time when all people in Qufu city were busy serving Qianlong who came to worship the ancestors, the chef in Kongfu was, however, helpless with the emperor' loss of appetite. He had no other choice but to use what was available — bean the sprouts which he stir-fried with some peppers. Fortunately, this dish with a light flavor was highly praised for neutralizing the grease that had been the reason for the emperor's loss of appetite. Since then, fried bean sprouts become a traditional dish of Kongfu Cuisine. The removal of root gives the bean sprout the shape of lilac and make them appear white. Cooked with shark fish, this dish became known as white lilac shaped shark fin given that those shinny white bean sprouts look like needles threading through white lilac being unique and elegant.

The mainstream values of Confucianism are considered

烧开，用慢火熬浓呈乳白色为奶汤。随后将鲍鱼切粗料，用本身的浓汤氽透。煮熟豌豆，开水氽滑子菇待用，清水洗菊花瓣。奶汤、鲍鱼、滑子菇、豌豆同煮烧开，调匀口味，用水淀粉勾稀芡，淋明油，撒上菊花瓣，入汤盅即成，带香菜碟一同上桌。奶汤鲍鱼羹一菜和五味而尚纯正，浓香清鲜，也显出了济南菜烫菜的特色。

换至胶东菜，厨师们追寻的五味调和则表现为保持主料的原味，如白汁酿鱼一菜。取鲜黄鱼，两面用坡刀法贴脊骨片斜刀七片，讲究薄均匀，肉不离骨。鱼头下颌及胸鳍处切开，两鳃壳向两边掰开，将刀拍鱼头使其趴下。猪瘦肉切成泥，加调料拌匀成馅。火腿、冬笋切薄象眼片待用。将肉馅抹入鱼片卷成卷，鱼腹朝下捏入鱼盆，两面鱼卷贴脊处竖起，每个鱼卷顶端按上一粒青豆，入笼用旺火蒸一刻钟取出。蒸鱼所得原汤加入炒锅，另加清汤、料酒，倒入火腿、冬笋片，沸后去沫，用湿淀粉勾薄芡，最后淋熟鸡油即成。成品色泽乳白，黄鱼原汁原味，汤鲜肉嫩。

除了儒家思想的要义在鲁菜中有所体现，孔子还针对饮食有过专门的论述，即食不厌精，脍不厌细。在孔子生活的年代，生产力水平有限，"食不厌精，脍不厌细"的真正含义，主要是针对祭祀。在物质生活较为丰富的今天，其含义

to be "rites" (*Li*), "virtue" (*De*) and "benevolence" (*Ren*), which require followers to distinguish right from wrong. The golden mean, which is defined as the felicitous middle between the extremes of excess and deficiency in Confucianism, embodies "Five tastes reconciliation" in Shandong Cuisine. To the chefs, it is an uphill task to accomplish the "Five tastes reconciliation" because it brought about difficulties and challenges that can solely be surmounted by using great wisdom and proficient skills. Masters of all schools that belong to Shandong Cuisine, however, show their special prowess in different aspects. For instance, masters of Jinan Cuisine are particularly adept at soups. The dish stewed abalone in milky soup is undoubtedly rated as the pearl of its type. Take abalone as its main ingredient and pettitoes, mushrooms, hen, coriander, chrysanthemum, pork lean, pig shin, and peas as the auxiliary ingredients. First, chop the pettitoes, wash the hen, pork lean and pig shin and put them in boiling water for several minutes. Next, move them to a soup bucket and boil until the colour of the soup turns milk-white; reduce the heat to simmer until it thickens. Second, chop the abalones and boil them until they are well done, then cook the peas till they are done and boil the mushroom for a while, then set all ingredients aside. Furthermore, wash clean daisy petals and set them aside. Put together the milk-white soup, abalone, mushrooms, and peas and boil them. Thereafter, mix several seasonings well, add water and starch to thicken, then boil. Finally, drizzle a little oil and strewn daisy petals and do not forget to serve it with coriander. This dish blends five tastes in perfect harmony thereby producing not only a savoury and fresh taste, but

also a unique Jinan style "hot pot."

As for Jiaodong Cuisine, the chefs' main focus is to maintain the original flavor as they pursue the "Five tastes reconciliation." A typical example is supposed to be the dish known as baizhi niangyu (braised yellow croaker). First, scale the fresh yellow croaker. Slice the fish along the spines into seven pieces of about 1.5cm thick from 45-degree angle on both sides. The thickness of the fillet needs to be well-distributed and they should remain attached to the fishbone. Cut the underjaw and the pectoral fin and ensure that the gill shells are apart on the two sides. Then pat the head down with a knife. Second, mash the pork lean, add seasonings and stir well to make fillings; cut ham and winter bamboo shoots into diamond-shaped pieces and set it aside. Then wrap the fillings into fillets and roll. After that, place the rolls into fish platter ensuring that the fish belly faces down. Erect the part of rolls near the spine and place a green bean on the top of each fish roll. Third, steam them over high heat for 1 minute then set them aside. Next, add the soup gained after steaming as well as clear soup and cooking wine to the wok and pour in ham and winter bamboo shoots and boil. Make starchy sauce with wet starch and drizzle cooking oil to finish. This dish, with a milky white color, tastes tender and delicious without losing the original flavor of the yellow croaker.

In addition to the essence of Confucianism, Confucius, opinion on diet, is also reflected in Shandong Cuisine, for example, "Eat no rice but is of the finest quality, nor meat but is finely minced; be very particular about one's food." This sentence, in the age when Confucius lived, was mainly a description of sacrifice given the

已转变为鲁菜当中对技法的追求。鲁菜发展至今，广泛应用的主要烹饪技法就多达数十种，如若细分，则更难以计数，相关名菜也数量众多。炖法之中有侉炖，菜肴如侉炖目鱼；熘法之中有糟熘，菜肴如糟熘梭鱼片；烹法之中有醋烹，菜肴如醋烹黄花鱼；糖粘法之中有拔丝，菜肴如拔丝八爪鱼；烧法之中有葱烧，菜肴如葱烧海参。在刀工方面，鲁菜厨师技艺也十分精湛，可在丝绸上切肉如丝，还有开创性的手拉绝活，如蝴蝶海参一菜。深入了解鲁菜之后会发现，技法是精髓，也是趣味。

山东属于温带季风气候，这一气候类型为亚洲所独有，因此与鲁菜更有可比性的是同属于这一气候的亚洲城市，如日本札幌、韩国首尔等。札幌是日本北海道地区的城市，在饮食上崇尚原味，如原味海胆、海胆蟹肉丼等。当地人还喜食鲱鱼，用盐稍腌，炭烤而食，追求的正是这种原生的鲜味。在韩国的首尔，海鲜也可用来制作泡菜，如包裹式泡菜，将栗子、梨、大枣、章鱼、鲍鱼、虾、松仁等以大白菜包裹，具有一般泡菜所没有的"豪华"，同时别具风味。这样的泡菜与其说是佐餐小菜，倒不如说是正餐菜肴，丰盛与风味毫不逊色。

走进鲁菜，才会发现，文化的韵味既在五味之内，又超

limited productivity level. While in an extensively materialized modern society, the meaning, however, has been transformed into the pursuit of more exquisite cooking skills in Shandong Cuisine. With the development of Shandong Cuisine, numerous cooking techniques have been developed among which dozens have been widely adopted. If these techniques are subclassified, it would even be more challenging to count. This also means there are many famous dishes as well. For example, "Kuadun" (first fry, and then stew) is a stew method with kuadun plaice as its representative dish ; "Zaoliu" (fry in wine sauce) is a kind of quick-fry method with zaoliu barracuda slice (fried barracuda slices in wine sauce) as its representative dish; "Cupeng" (boil with vinegar) is a method of boiling with cupeng yellow croaker (sweet and sour yellow croaker) as its representative dish; "Basi" (hot candied) is a kind of sugar adhibit method with basi octopus (hot candied octopus) as its representative dish; "Congshao" (braise with scallions) is a kind of braise method with congshao sea cucumber (Braised sea cucumber with scallions) as its representative dish. Masters have such high-level knife maneuvering skills that they can slice meat on silk. In addition, they are the owners of other groundbreaking forte as well, which is somewhat, reflected through famous dishes such as butterfly-shaped sea cucumber. The cooking techniques are seen as both quintessence and fun of Shandong Cuisine with one getting to know it better.

Shandong has a monsoon climate that is unique to Asia. Therefore, what is comparable to Shandong Cuisine is the cuisine of Asian cities with the same climate. They include the delicacies of

越五味。纵览典籍，"品"本指事物的种类和等级，"格"本指人的风格度量或事物的法式标准，品与格并用，有了品级、美感、标准的含义。回顾鲁菜绵延发展的历程，味道中包含着先人们的智慧，品级并非优劣，美感拘无定式，标准不渝经验，这种大智慧，是美食文化，是儒家思想，更是中国文化。植根于中国文化的鲁菜海鲜美食，会继续绵延它的品格。

Sapporo, Japan and Seoul, Korea. Sapporo, a city in the Hokkaido region of Japan, boasts an ability to pursue the original flavor of food, including original sea urchin and Donburi of sea urchin and crab. The grilled herring fish, which is slightly salted, is also the locals' favorite for its original fresh taste. While in Seoul, South Korea, seafood can also be used to make Kimchi, such as wrap-style Kimchi which, after wrapping chestnuts, pears, jujubes, octopus, abalone, shrimp, pine nuts, etc. with cabbages, has the "luxury" and flavour that is different from ordinary Kimchi. Such Kimchi is not so much a side dish, but rather a luscious dinner, sumptuous and full of flavor.

As you know more about Lu Cuisine, you will realize the charm of the culture embodied in and beyond the five flavors. Reading through all the classics, we can find that the original meaning of *Pin* is categories and levels of things while *Ge* refers to magnanimity and behavioral pattern of a person or mandatory standard of things. A combination of *Pin* and *Ge* endows Shandong Cuisine with nobility, beauty and standard. It is solely by reviewing the historical development of Shandong Cuisine can we recognize the wisdom of our ancients. Experienced chefs of Shandong Cuisine created such artistic dishes which are of high quality with elegance and popularity. Shandong Cuisine is not only a type of cuisine culture, but also an embodiment of Confucianism, or even, a representative of Chinese culture. The seafood delicacies of Shandong Cuisine rooted in Chinese culture will continue to be inherited.

本味，吴地的平和

第四章

Jiangsu Cuisine:
Savory Taste

　　康熙六年，江南省一分为二，其中一省取江宁、苏州两府首字，名江苏。清朝时江苏省所辖地域与今天相差不多，若从最初的勾吴古国算起，则辖域多有变化，但不可否认的是，吴地的中心，始终在今天江苏境内。古时吴地，今时江苏，变迁是历史，也是滋味，我们要讲的就是历史中的那份滋味。

　　江苏有着全国八大菜系之一的苏菜，美食佳肴不胜枚举，但要谈起滋味，就要先讲一个故事。相传彭祖之子夕丁好捕鱼，彭祖恐其溺水，便禁止其捕鱼。一日，夕丁得一鱼，央求母亲烹制，母亲正炖羊肉，遂藏鱼于羊肉中同烹，未料成菜鲜美异常。"羊方藏鱼"成就了"鲜"，也造就了江苏菜的精髓。江苏菜追求本味，讲究平和，在本味与平和之中，"鲜"是最亮丽的色彩，而这份"鲜"的决定因素就在于食材。

Jiangsu became an independent Chinese province in 1667. The name was derived from the prefixes of Jiangning and Suzhou, the names of the two most important prefectures within the province at that time. In the Qing Dynasty, the area under the jurisdiction of Jiangsu Province was much the same as today. Despite the changes in its jurisdiction since the ancient State of Wu, Jiangsu has remained the center of the Wu area. It is the history and the taste that has transitioned from the State of Wu to Jiangsu today. In this chapter, we are going to talk about the history of Jiangsu Cuisine.

As a food paradise, Jiangsu boasts numerous delicacies. And Jingsu Cuisine, also known as Su Cuisine, is rated as one of the Eight Major Cuisines of China. It would be necessary to narrate a story prior to delving into these "tastes." It is said that Xiding, the youngest son of Pengzu, a legendary figure in China who lived for a long time, had a deep love for fishing. Unfortunately, his father forbade him from fishing because he feared that his son may drown. One day, Xiding caught a fish and he pleaded with his mother who was stewing mutton to cook it for him. His mother then had no choice but to wrap the fish with mutton and cook together. The dish was found extremely savory, so it was named steamed fish wrapped in mutton. The Chinese character fish (鱼) and mutton(羊) combining together made a new character "savory" (鲜), which gradually became a key word in Chinese food culture. In general, care is always taken to preserve the original flavor in Jiangsu Cuisine. With soft texture, the taste of Jiangsu Cuisine, however, is far from insipid because "savory" illuminates its unique beauty. Furthermore, the decisive "savoury" factor lies in the ingredients.

中国七大水系中的长江、淮河横贯江苏，五大湖泊中的太湖、洪泽湖也位于江苏，与此同时，江苏的海岸线长达近千公里，为苏菜的发展提供了得天独厚的条件。这正是古籍所载的："吴地产鱼，吴人善治食品，其来久矣"。江苏菜有两千多年的历史，但盛于唐宋，南宋时，与浙菜同为"南食"的两大台柱。根据地域和特色的差异，苏菜主要可以划分为四大类：徐海菜、淮扬菜、南京菜和苏锡菜，我们从北到南依次讲起。

从地理位置上看，徐海菜所属苏北地区，靠近山东，故菜肴的口味、色调多近齐鲁风味。徐海菜又以徐州和连云港菜为主，徐州素有"五省通衢"之称，而连云港则是著名的海港，海产品十分丰富，海鲜入菜顺理成章。徐海菜烹调技艺多用煮、煎、炸等，譬如清代宫廷菜凤尾大虾。此菜主料为连云港所产海洋大虾，烹饪时取鲜虾十只，去头壳，留尾。将虾从腹部片开，用葱姜、盐、料酒腌制入味。另将鸡蛋、淀粉调成糊。热锅下油，待油烧至五成热时，用虾挂蛋糊入油锅炸成凤尾状捞出，再入七成油温复炸，装盘即成。这道菜用料不多，但口感外酥里嫩，海虾的味道鲜香浓郁，造型美如凤尾，因此成为宫廷菜。

因连云港盛产墨鱼，故以墨鱼为主料的菜肴也颇多，

As two of China's seven major rivers, the Yangtze River and Huaihe River that pass Jiangsu Province created a well-developed water system. Likewise Taihu Lake and Hongze Lake which are among the five major lakes are also located in Jiangsu. At the same time, the coastline of Jiangsu stretches nearly a thousand kilometers and offers unique conditions for the development of Su Cuisine. As recorded in ancient books: "Wu area is rich in marine products, and people here have always been good at cooking fish." Although Jiangsu Cuisine has a long history of more than two thousand years, it was in the Tang and Song Dynasties that it reached its heyday. In the Southern Song Dynasty, Jiangsu Cuisine, together with Zhejiang Cuisine, was considered the two mainstays of "Southern Cuisine." Based on the differences in regions and characteristics, Jiangsu Cuisine can mainly be divided into four categories: Xuhai Cuisine, Huaiyang Cuisine, Nanjing Cuisine and Suxi Cuisine. We will discuss about these cuisines beginning from the north to the south.

Geographically, the northern Jiangsu area is the birthplace of Xuhai Cuisine. This area connects Jiangsu and Shandong provinces and because of this proximity, the taste and colour of dishes here bear significant resemblance to the dishes of Qilu (Shandong) area. Xuhai Cuisine is mainly composed of Xuzhou Cuisine and Lianyungang Cuisine. Xuzhou city has long been known as "the crossroad of five provinces" while Lianyungang City is a famous harbor abundant with seafood products. It stands to reason that seafood is the main ingredient in this locality. Xuhai Cuisine is cooked by boiling, frying and deep-frying. For example, the dish prawns in phoenix's tail shape whose main raw material is sea

如做工讲究的席面菜玉莲蟹黄墨鱼球。将墨鱼肉斩成泥茸，加鸡蛋清、猪肥膘、马蹄、调料一起搅上劲，包入蟹黄，粘面包糠，炸至金黄色置盘中，最后再以小墨鱼卷成马蹄莲状围边。

再如普通人家的爆乌花，取鲜墨鱼肉剖荔枝花刀块，入沸水略氽，沥水，热锅下油，待油热至六成，下葱姜蒜末爆香，入乌花，加玉兰片、木耳、青豆、香菜等配料，以绍酒、白糖、味精调味，入水淀粉勾芡，撒胡椒粉出锅即可。此外还有油爆海螺片、红烧沙光鱼等种种地方特色菜肴。虽说原材料各不相同，却是同样鲜香浓郁。

脚步再向南，就到了淮扬地区。淮扬菜始于春秋，兴于隋唐，盛于明清，素有"东南第一佳味，天下之至美"的美称。其选料严谨，多以水产为主；制作精细、刀工讲究；追求本味、清鲜平和。在江苏淮安市东郊有一个小镇，名为朱桥镇，该镇的甲鱼菜肴别具特色，甲鱼羹堪称其杰出代表。此菜选用地产野生甲鱼和水发干贝为主料，制作时先将甲鱼宰杀洗净，入沸水煮至八成熟，拆骨取肉切成小丁，然后于锅中加入底油，入葱、姜，放入干贝、鸡汤、笋丁、甲鱼肉、猪皮冻，大火烧至汤汁浓稠，加盐、味精等调味装入分盅，撒香菜即成。河鲜与海鲜的味道融合在一起，香气四溢，令人回

prawns was an imperial dish in the Qing Dynasty. In order to prepare it, first, select ten fresh prawns and peel off their shells (heads). Make sure to keep their tails. Then slice each prawn along the abdomen and then marinate them with salt, scallions, ginger and cooking wine. Next, mix eggs and starch to make paste. Second, put the wok over the fire and pour oil. Then fry the sliced prawns with egg paste in 125-degree oil until they shape like the phoenix's tails. Finally, fry them again in hot oil with a temperature of 175 degrees and transfer to a plate to finish. This dish tastes crunchy outside and soft inside. Savory flavor along with an attractive phoenix's tail-shaped appearance makes it a dish for imperial palace.

Furthermore, there are numerous dishes with cuttlefish as their main ingredient given that cuttlefish abounds in Lianyun'gang. For example, the dish callalily-shaped cuttle balls with crab cream is an exquisite banquet dish. In order to cook it, begin by mashing cuttlefish before adding egg white, pork fat, calla and seasoning to mix well. Next, wrap crab cream, coat with breadcrumb and fry until it turns golden brown. Finally, roll the little cuttlefish into the shape of calla and arrange them around the plate.

Another dish which often appears in ordinary homes is quick-fried cuttlefish. To cook it, begin by carving fresh cuttlefish in the shape of litchi. Next, blanch in boiling water for a little while, then drain the water. Second, add scallions, ginger and chopped garlic then quick-fry in hot oil with a temperature of 150 degrees. Add cuttlefish and ingredients like water-soaked bamboo slice, agaric, green beans and coriander as well as seasonings like Shaoxing wine, sugar, and MSG that help to blend flavors. Finally, dress with

味无穷。

除了河鲜与海鲜的并用，淮扬地区还有一种洄游性刀鱼。此鱼早春肉质肥美，然一过清明，刺变硬，肉转老，故有谚云："春有刀鲚夏有鲥"。淮扬名菜双皮刀鱼便是以此鱼为主料。双皮刀鱼之出名除了味美，还在于其刀法之讲究。制作时，需将刀鱼刮鳞，去鳃、鳍，在肛门处横划一刀，用竹筷从鳃口插入鱼腹，绞去内脏，切掉鱼尾，然后在鱼背处用刀沿脊骨两侧剖开，去掉脊骨，再将鱼皮朝下平铺，用刀背轻捶鱼肉，使细刺粘在鱼皮上，最后用刀面沾水，刮下两面鱼肉。此过程中要求既不刮破鱼皮，又要使鱼皮上留少量鱼肉，还要去掉鱼皮上的鱼刺，足见刀法之精细高超。刀工之后，需将刮下的刀鱼肉剁成鱼茸，与猪熟肥膘茸、鸡蛋清、精盐、味精、绍酒和适量清水搅匀，平铺在刀鱼皮的肉面上，再将鱼皮合上成鱼原状，在合口处沾上火腿末、香菜末，然后整齐入盘。再将火腿片、春笋片、冬菇片相间铺在鱼身上，加葱姜、绍酒、精盐，上笼蒸熟取出，拣去葱、姜，滗去汤汁。上桌之前，还需以清鸡汤浇在鱼上提鲜。此菜成品鱼形完整无骨，鱼肉保持本味，汤汁留有鸡的鲜味，交融之中让人感受到淮扬菜的无限美妙。

starchy sauce and pepper to finish. In addition, there are a variety of local specialties such as fried conch slices, braised javelin goby. Different as the raw materials are, they taste equally savory and have a strong aromatic smell.

Let's go further south to Huaiyang region. Huaiyang Cuisine first appeared in the Spring and Autumn Period and developed in the Sui and Tang Dynasties. Its day of glory began during the Ming and Qing Dynasties when it was lauded as "The best delicacy in the southeast of Yangtze river and second to none in the world." Its outstanding characteristics owe much to the meticulous selection of ingredients comprising essentially of freshwater aquatic products. With exquisite cutting and shaping techniques, Jiangsu Cuisine seeks to maintain its original savory and smooth qualities. In the eastern suburbs of Huai'an City, Jiangsu Province, Zhuqiao Town—a small town well-known for its unique turtle specialties is located. Wild turtle soup whose main ingredients are wild turtles and soaked scallops appears as an outstanding representative. To prepare it, first, slaughter wild turtles and wash them clean. Then boil them until they are about 80% done. Then take out their bones and cut them into small cubes. Second, put scallions, gingers, scallops, chicken soup, cubed bamboo shoots, turtles and pigskin into the heated wok with oil and cook with high heat until the mix thickens. Finally, add seasonings like salt and MSG. Arrange them in separate tureens and dress with coriander to finish. A perfect blend of the flavor of seafood and freshwater fishes gives the dish a long-lasting savory taste and smell.

Besides a combination of seafood and freshwater fishes, there is also a kind of migratory ribbonfish in Huaiyang area. In early

说吴地的中心一直在江苏，一是因为苏州，再一是因为南京。南京人杰地灵，南京菜也独树一帜，至明清成流派，在民国期间则达到巅峰，蜚声中外。因为是六朝古都，文人墨客云集，南京菜讲究细致精美，具有名贵典雅、华美大气的古都风范。翡翠虾饼便是南京菜中别具一格的花色菜。取初春蚕豆制绿泥，将鲜虾仁斩茸，放熟肥膘粒和鸡蛋清，加精盐、绍酒、味精、干淀粉搅拌均匀成虾缔，捏成圆形小饼数只，中间以火腿末点缀。热锅下油，将小饼滑入温油中边贴、边氽、边舀热油浇在虾饼上，待中间鼓起，色呈碧绿时滗去余油，干贴片刻，最后整齐地排放盘中即可。食用时可配以蘸碟。这道菜色如翡翠，底壳酥脆，口味中既有蚕豆的清香，也有虾肉的鲜美。

　　清末，南京作为口岸开放后，受到西方文化影响颇多，这一点也明显体现在菜肴中，譬如这道金丝香鳕球：将土豆切细丝炸至金黄色，深海银鳕鱼切成小方块用精盐和味精腌渍后，拍上干淀粉炸脆；在卡夫奇妙酱中加入适量番茄沙司，拌匀；将银鳕鱼块沾匀酱后裹满炸好的土豆丝，整齐摆放盘中即成。其成菜色泽金黄，香甜微酸，风味独特。此菜以银鳕鱼和土豆为主料，佐以西餐中的卡夫奇妙酱，是一道中西合璧的佳肴，西式菜肴对南京菜的影响可见一斑。

spring, this ribbonfish is most rich and most tender. However, the fish becomes rough just after early April. An old proverb therefore goes: "We have ribbonfish in spring and ilish in summer." Taking the fish as its main raw material, the famous Huaiyang dish double skin ribbonfish is well-known not only for its delectable taste, but also for its exquisite cutting skills. It is cooked by first, scaling the fish and removing the fins and gills. Then the fish is crosscut in the anus and a bamboo chopstick is inserted from the gill to the belly in order to remove abdominal organs and the fish tail. After that, the fish is cut along both sides of the backbone to remove the spine and its skin is spread down and is flapped with a knife back in order to make the tiny spines stick to the fish skin. Finally, scrape fish with wet knives. The whole process, which requires not only to keep fish skin intact, but also to leave small amounts of fish on the skin and remove the tiny spines, is a demonstration of the exquisite and skillful cutting techniques. After that, the fish is mushed and it is mixed with mashed and cooked pork fat, white egg, salt, MSG, and Shaoxing wine. Then it is stirred well after adding clean water. Thereafter, the fish is laid down ensuring that the fish which is still stuck to the fish skin is on the downside, then the two halves are put together in the original shape of the fish. Next, chopped ham and coriander are put into the incisions of the fish then moved to the plate. Then ham, spring bamboo shoots and snake butter are sliced and coated over the fish. Furthermore, scallions, ginger, salt are added and steamed until they are well done. Finally, the scallions and ginger are removed while the broth is decanted. Remember to pour clear chicken soup on the fish before enjoying. The dish, which looks like the complete shape

of a fish but has no bones, is a perfect fusion of the savory taste of chicken soup demonstrating the fantastic Huaiyang flavour.

As we have mentioned above, Jiangsu has long been the center of the Wu area. It is undoubtedly because of both the existence of Suzhou, and Nanjing—a remarkable place where many talents were gathered. Nanjing dishes have won fame both at home and abroad for their unique style which was developed during the Ming and Qing Dynasties and they gained enormous popularity during the Republic of China. Nanjing is the capital of six dynasties in Chinese history and a gathering place of the men of letters. Thus, Nanjing Cuisine is famous for its delicacy, elegance and splendidness, with characteristics that greatly exuded the charm of an ancient capital. Emerald shrimp cake is a unique bright-colored dish which is cooked by first pureeing spring broad beans, then mashing the fresh shrimps. Next, cooked pork fat, white egg, salt, Shaoxing wine, MSG and dried starch are added and stirred well to produce shrimp paste. The paste is then shaped into several couples of small round cakes and it is garnished with ham in the middle. Shrimp cakes are slowly slid in the heated wok with oil along the surface in order to bake and, at the same time, blanch them. At the same time hot oil is poured over shrimp cakes until their middle part is bulged. Finally, the rest oil is decanted and it is baked for a while. The shrimp cakes are neatly arranged in a plate and they may be served with dipping sauce. The dish, with an emerald color and crisp bottom, fuses the crispness of the broad beans and the freshness of the shrimps in a perfect way.

Nanjing has been greatly influenced by western culture since it opened as a treaty port at the end of the Qing Dynasty. The

脚步停在苏锡一带，驰名中外的苏州园林坐落在这里，传承千年的烹饪技艺也诞生在这里。《史记·刺客列传》曾记载专诸刺杀王僚的故事，而专诸得以见到王僚，便是因为其习得了"炙鱼"之法，为王僚侍菜。由此可知苏锡菜烹鱼的悠久历史。在众多鱼肴之中，又以一道清蒸鲥鱼最为出名。鲥鱼每年五月由海溯江产卵，肥美异常。旧时，鲥鱼曾是敬奉皇帝的"御膳"珍馐，清朝康熙皇帝就曾下令从扬子江"飞递时鲜，以供上御"。关于这道菜，还有一个动人的故事。东汉时，有个叫严光的人，曾助汉光武帝刘秀建朝立业，后刘秀称帝，严光却隐居江畔，刘秀多次遣使说服，才将他接到京城。为说动严光出任相助，刘秀甚至与其"共卧"，然而严光津津有味地讲他垂钓鲥鱼清蒸下酒的美味，使得刘秀亦不觉口中生津。严光终以难舍鲥鱼美味，婉言谢绝了做官，成为佳话。

制作这道菜时，需将鲥鱼挖去鳃，去内脏，沿脊骨剖成两片，不去鳞，用洁布吸干水。将猪网油洗净，晾干；鱼入沸水烫去腥味后，鱼鳞朝上放入盘中，将火腿片、香菇片、笋尖相间铺放在鱼身上，再加熟猪油、白糖、精盐、虾子、绍酒、鸡清汤，盖上猪网油，放上葱段、姜片，上笼蒸熟取出，拣去葱、姜，剥掉网油，滗出汤汁，加白胡椒粉调和，

influence is apparent in Nanjing dishes such as golden silver pout ball which is cooked by first slicing potatoes and frying them until they turn golden brown. After cubing silver pout and marinating with salt and MSG, coat silver pout with dried starch and fry until they turn crisp. Second, moderate tomato sauce is added into the miracle whip and stirred well. Finally, the cubed sliver pout is equably stained with sauce and it is coated with fried potato slices which are then arranged neatly in a plate. The dish has a completely unique flavor, a golden-brown color, and it tastes moderately sweet and a little sour. The main raw materials of the dish are sliver pout and potatoes combined with western miracle whip. Therefore, it deserves to be a sino-western dish bearing testimony of the Western Cuisine's influence on Nanjing dishes.

A visit to Suzhou-Wuxi area should definitely include some time in the world-famous Suzhou Gardens to soak up the typical lifestyle. It is here that the millennium-aged cooking techniques were born. Records of *the Grand Historian: Biographies of Assassins* has the story of Zhuanzhu (an assassin in the Spring and Autumn period) killing Liao (king of the State of Wu). Zhuanzhu was able to approach King Liao because of his mastery of a special cooking technique to "roast fishes." He then became an understrapper who served dishes for Liao. From the story, we can see that Suzhou-Wuxi area is known for its tradition of cooking fish. One of its most famous dishes is steamed ilish. Ilish spawns in the Haishuo River every May when it is at its tenderest. As a tribute for the imperial palace, the dish used to be ordered by Emperor Kangxi to "fast deliver to the capital." There is also an interesting story about the dish. It is said that in the Eastern

海错食单

The Stories of
the Chinese Seafood

Han Dynasty, a man named Yan Guang helped Emperor Gungwu Liu Xiu (5BC–57BC) to establish the dynasty and then left to spend his life in seclusion. Liu Xiu persuaded Yan Guang in every possible way to be his prime minister, Yan Guang, however, only talked with great enthusiasm about his experience of fishing and cooking the ilish as well as its wonderful taste when served with wine. This description made Liu Xiu's mouth water as well. Finally, Yan Guang found it impossible to abandon the delicious ilish, so he declined the Emperor's invitation to be an official. This became a favorite tale for later generations.

The cooking method is not complicated: First, the gills and abdominal organs are removed and the fish is cut in half. The scales are drained with a clean cloth and they are set aside. Then the intestinal pork fat is washed clean and drained. Next, the fish is blanched over boiling water and then arranged on plate. The sliced ham, mushrooms, bamboo shoots are then spread on the fish with each of these ingredients alternating with each other. Thereafter, lard, white sugar, salt, shrimps, Shaoxing wine, clear chicken soup are added unto the plate and are covered with intestinal pork fat. Then chopped scallions and sliced gingers are added and steam until they are done. Finally, scallions, gingers and intestinal pork fat are removed, the broth is decanted, moderate pepper is added and mixed well and the broth is poured on the fish. Finish by dressing with coriander. Remember to serve with bruised ginger and vinegar. The dish has a bright and beautiful color and its taste is tender, delicious and refreshing. It is the true reflection of Suzhou dishes to pursue originality of flavor and of light taste.

再浇在鱼身上，放上香菜即成。上菜时带姜末、香醋碟，以供蘸食。这道菜色泽美观，肥嫩鲜美，爽口不腻，正是苏菜追求本味、讲究平和的体现。

苏菜的本味与平和，在环境类似的世界其他地域的海鲜美食中也有所体现。例如在佛罗里达，有独特的石蟹，其蟹钳一大一小。吃石蟹，也十分独特，只吃蟹钳，蒸或煮之后，去除蟹钳的肉，直接蘸食，吃的是原味与香甜。在摩洛哥，沙丁鱼是出口量最大的鱼类，在市场上可以买到各种沙丁鱼酱。当地人喜欢将酱涂在油炸沙丁鱼上食用，这种吃法称得上是原汁原味。在南半球的圣地亚哥，面包蟹汤则很受欢迎。而在悉尼，岩牡蛎只需配以柠檬汁和葡萄酒，嫩滑鲜美就可以充斥唇齿之间。不同地域，对本味与平和的追求有着不同的实现方式，但"鲜"却是共通的。

"江南好，风景旧曾谙。"诗人白居易在落笔之时，所怀念的江南，想必有美景，也有美食。漫步在海风徐徐的滩涂上，古时吴地，今时江苏，时间流淌，海风的味道未曾改变，苏菜的本味也将延续，一如最初的平和。

The original and light flavour, of course, is by no means exclusive to Jiangsu Cuisine, but can be embodied on other seafood delicacies in the world. For example, in Florida, there is a unique kind of king crab with one big claw and one small claw. A favorite local treat is also unique for they eat only the crab claws. After steaming or boiling, diners separate the crab meat from the shells and serve with seasonings so as to preserve the original flavor and sweet tastes. In addition, as the largest sardine exporter in the world, Morocco has a variety of sardine sauces available on the market. Locals opt to spread the sauces on fried sardines to enjoy the sardine flavour itself. In the southern hemisphere lies Santiago where Brown Crab Soup is popular while in Sydney, serving delicate oysters with lemon juice and wine makes them simple enough for a lip-smacking meal. Although people in different regions show their special prowess in the pursuit of the original flavour, "savory" is always something in common. It meets at the tip of their tongues, and within their hearts.

"Fair southern shore, with scenes I adore." When the poet Bai Juyi (AD772－846) wrote this line, the scene of Jiangnan (the southern shore) unveiled and memories came gushing to his mind. These memories probably included beautiful sceneries and various delicacies. Taking a stroll on the beach where the gentle breeze is blowing, it is possible to sigh at how the fleeting time changed the ancient Wu area into Jiangsu as we know it today. However, as the "flavor" of the sea can never be changed either now or in the future, Jiangsu Cuisine will also continue to preserve its original flavor— gentle as what it was at the beginning.

内蕴，两浙的情致

第五章

Chapter

Five

Zhejiang Cuisine:
Inner Beauty

自古以来，以钱塘江为界，江水以南为浙东、以北为浙西，两浙合为浙江。脚步踏入两浙，登天台山，赏西湖水，观钱塘潮，游人皆如醉。大自然造就的千里画廊，正是东南形胜，吸引着无数文人墨客、商贾旅人。人们聚坐而食，离别以酒，又造就了饮食文化。物产、风景、文化在这里碰撞，造就的是两浙的内蕴与情致。

浙江有着长达六千四百余公里的海岸线，渔场密布，盛产海味。《皇帝内经》有曰："东方之域，天地所始生也，渔盐之地，海滨傍水，其民食鱼嗜咸，皆安其处，美其食"。由此可见，浙江以海味入食已有数千年的历史。春秋时期，越国定都会稽（今绍兴市），经济繁盛，极大促进了钱塘江流域烹饪的发展。南北朝以后，江南数百年免于战乱，隋唐又开通京杭大运河，对外贸易日益频繁，当时的宫廷菜肴和民间饮食等烹饪技艺得到了长足发展。至南宋时期，宋室南

From the ancient times, the Qiantang River divided its surrounding regions into two: the south of the river known as Zhedong and the north known as Zhexi. Later, the two regions were combined to become Zhejiang Province. Visitors to Zhejiang Province can scale the Tiantai Mountain, enjoy the West Lake, and appreciate the Qiantang bore.❶ Every year, there is a continuous stream of visitors who come repeatedly to Zhejiang to indulge in its beauty. The thousand miles of scenic gallery is so charming that it has attracted countless literati, poets, merchants, and tourists at all times. People gather, talk and eat, or drink when bidding fare well thereby creating a special food culture. The collision of local products, sceneries and cultures here add to the beauty and deepen the impact of Zhejiang Cuisine.

With a coastline of over 6,400 kilometers, Zhejiang Province boasts lots of fishing grounds which are rich in seafood. *The Inner Canon of Huangdi* (*Huangdi Neijing*), an ancient Chinese medicine classic, states, "There is a place near the sea in the east, abundant in fish and salt, where people like eating fish and salty food. They are living a life of peace and content." It is perceptible that Zhejiang people have been eating seafood for thousands of years. During the Spring and Autumn Period, the Yue State chose Kuaiji (present day Shaoxing City, Zhejiang Province) as the capital. The local economy of the city was subsequently developed, greatly promoting the development of cooking in the Qiantang River area. After the Northern and Southern Dynasties, Zhejiang and its neighboring regions did not engage in any wars for hundreds of years. Later the Sui and Tang dynasties built the Beijing-Hangzhou Grand Canal

❶

Tiantai Mountain, West Lake and the Qiantang bore are all famous scenic spots in Zhejiang Province.

and ever since, commercial trade has become increasingly frequent. At the same time, the cooking skills displayed in the preparation of imperial dishes and folk diets were greatly developed. In the Southern Song Dynasty, the royal families immigrated to the south because of wars and chose. Hangzhou (the present-day capital city of Zhejiang Province) as the capital of the whole country. Immigration also enabled the North-South diets and culinary culture to integrate, bringing Zhejiang Cuisine to its apogee with an endless variety of famous dishes. In the history of the development of Zhejiang Cuisine, culture and food are closely linked. The implication of the culture tints each dish while the appeal of the food is also reflected in the culture. Therefore, Zhejiang dishes are often connected with all sorts of myths, historical stories, anecdotes, and literary articles. While savoring the deliciousness of Zhejiang Cuisine, people can simultaneously enjoy a feast of culture too.

The culture of mankind begins with primitive religions and myths. So, with regard to seafood dishes in Zhejiang Cuisine, we will start with a myth. Ancient Chinese books told the story of a razor clam that was once a worm. After hearing of the legendary experience of the Monkey King, it became so ambitious that it called itself the Sea King. Because of the previous turmoil caused by the Monkey King, the Dragon King was very afraid of the action of the clam. But realizing that it was just a razor clam, he was so angry, so he ordered a crusade against the razor clam. Unfortunately, this Sea King didn't have any magic power as the Monkey King. It was so terrified that it committed suicide by hanging itself. However, the suicide did not work as planned, so the clam did not die. It

渡，建都杭州，更是把北方的烹饪文化带到了浙江，使南北饮食广泛融合，浙菜达到极盛，名菜名馔层出不穷。在浙菜发展的历史中，文化与饮食紧密相连，内蕴喻于菜肴之内，情致显于文化之中，浙菜菜肴往往有与之关联的神话、史话、趣闻、书文等。人们在享受浙菜美味的同时，也享受了文化的盛宴。

人类的文化始于原始宗教和神话，关于浙菜中的海鲜菜肴，我们也从神话讲起。古书曾有这样一则神话：蛏原为无名海虫，妄图与齐天大圣攀比，自封海圣。因齐天大圣曾大闹龙宫，早把龙王吓得不轻，如今又冒出个海圣，龙王勃然大怒，下令讨伐。可惜这海圣不比齐天大圣那般神通广大，竟吓得上吊自缢。可它用的竟是朽绳，上吊未成，反而一头扎进泥滩，从此被讥为虫圣了。蛏由此得名。浙江台州沿海产蛏甚多，个大、壳薄、肉嫩、味鲜，明清时被列为贡品。浙菜中蛏的食用方法颇多，三丝拌蛏最为有名。此菜以蛏肉为主料，辅以多种配料凉拌而成，为夏季美食。制作时先将蛏肉入水煮熟，再将青椒丝入沸水中略焯捞出，并同香菇丝一起用煮蛏子的原汤浸渍两分钟，捞出冷却。将以上食材加绍酒、精盐、味精适量拌匀，入盘摆成葵花形，四周镶上黄瓜花，淋芝麻油、撒熟火腿丝即可。最后佐以用嫩姜末、醋、

could only plunge into the mud flat and was sneeringly called the Worm King since then. That is how the razor clam got its Chinese name 蛏 (a Chinese character composed of the word "worm" and "king").

There are many razor clams in the coastal areas of Taizhou City, Zhejiang Province. They are large, thin-shelled, and they also have tender and fresh meat. They were even listed as tributes during the Ming and Qing dynasties. This kind of clam is part of many dishes including the famous razor clams with three slices. This dish is mainly made from razor clam flesh and it is supplemented with a variety of other ingredients. It is a typical summer dish. To cook it, boil the flesh of the clam well, then put sliced green peppers in boiling water for a little while. After that, dip it together with sliced mushrooms into the soup of boiled clams for two minutes, then remove it and allow it to cool down. Second, add Shaoxing wine, salt, and MSG and mix well. Shape it into a sunflower and put it on a plate. After that, put cucumber flowers around the sunflower, drizzle a touch of sesame oil and dress it with sliced ham. Finally, serve the dish with a sauce of bruised tender ginger, vinegar, soy sauce, and pepper. The finished dish is tender and delicious to the taste and beautiful to the eye.

Human memory of taste is often profound and long-lasting, while the taste of food in history is always related to interesting stories. For people living at the seaside, "Xiang" (鲞) is an indispensable food commonly known as dried fish. What we are going to talk about is a famous dish in Ningbo City, Zhejiang Province, and it is called the *new-style dried eel*. This dish is a must-

酱油、胡椒粉调成的味碟上席。此菜成品蛏肉细嫩，味道鲜美，且菜品形态精美。

人们对于味觉的记忆往往深刻而久远，而历史当中食物的味道，则是有趣味的史话。对于生活在海滨人们而言，"鲞"是不可缺少的食物，也就是人们俗称的鱼干。我们要讲的是一道宁波地区的风味名菜——新风鳗鲞。这道菜是当地渔家春节时餐桌上的必备佳肴。相传春秋末期，吴王夫差与越国交战，带兵攻陷越地鄞邑（即宁波地区），御厨在五鼎食中，除牛肉、羊肉、麋肉、猪肉外，取当地鳗鲞，代替鲜鱼作菜。吴王食后，赞誉其味美，乃鲤鱼、鲫鱼所不能及，从此鳗鲞身价倍增。

新风鳗鲞是宁波人在冬令时节及过春节时制作的一种鱼鲞，略为风干，即可食用。制作这道菜时需先将鳗鲞洗净斩段，加绍酒、葱结、姜片，上笼蒸熟，然后去皮去骨，顺丝纹撕成小条，置碗内，入绍酒、精盐、味精、白糖、芝麻油拌匀，装盘即成。"鲞"以鱼为名，以鳗鱼制作的"鲞"叫做"鳗鲞"，而以黄鱼制作的"鲞"则叫"黄鱼鲞"，又称"白鲞"，例如绍兴名菜白鲞扣鸡，此菜以熟鸡肉脯、鸡翅膀、白鲞加入调料上笼蒸制而成，其成菜白鲞、越鸡两味相渗，鲜香浓郁，风味独特。

eat for the local fishermen during the Spring Festival. It is said that at the end of the Spring and Autumn Period, Fuchai, King of the Wu State, led a war to fight against the Yue State and his soldiers captured Yinyi (today's Ningbo area). Due to the food shortage, the royal chef then added local dried eel instead of fresh fish into beef, lamb, elk meat, pork in the meal he was cooking. After the King enjoyed the food, he profoundly praised its deliciousness, which made the taste of crucian and carp inferior in comparison. From that time, there has been an increase in the price of dried eel.

Ningbo people usually make this dish during winter and at the Spring Festival. The meal is cooked by first washing the dried eel. Then, cut it into chunks and add Shaoxing wine, scallion knots, ginger slices to the fish and allow to steam it well. Next, remove the skin and the bones of the steamed fish chunks, tear the meat into small strips and put them into a bowl. Finally, mix well with Shaoxing wine, salt, MSG, sugar, and sesame oil. The dried eel is called Eel Xiang (鳗鲞), while the dried yellow croaker is called Yellow Croaker Xiang (黄鱼鲞) or White Xiang (白鲞). For example, an well-known representative Shaoxing dish is called steamed white dried croaker with chicken. This dish is made from cooked chicken breasts, chicken wings, and dried yellow croaker with seasonings. The integration of dried fish and chicken gives the dish a unique taste.

Yellow croakers are among the top-grade sea fishes that are eaten. Ningbo is rich in yellow croakers, so there are many dishes of them. In addition to air-drying a yellow croaker, the locals also use it in various soups including the most typical yellow croaker

黄鱼是海鱼中的上品。浙江宁波地区盛产黄鱼，因此颇多以黄鱼为主料的菜品。在吃法上除了将黄鱼制成鱼干以外，还可做成黄鱼羹，咸菜大汤黄鱼便是其中之一。咸菜宁波人称咸鸡。民间传说明洪武末年新任明州（今宁波）知府王珏，为政清廉，常以糟水和咸菜宴请当地官豪，席间即吟《咸鸡赋》："一想百姓苦，菜味似鸡鲜"借以助兴。自此，"咸鸡"宴传为佳话，咸菜也由此被誉为咸鸡。咸菜用雪里蕻芥菜制成的为上品，在宁波是居家常备之菜，当地民间有句俚语："三天勿喝咸菜汤，两脚有点酸汪汪。"制作此菜时取新鲜大黄鱼一尾，宰杀洗净。炒锅烧热，下猪油烧至八成热，将黄鱼下锅稍煎，烹酒，加葱姜、鲜汤、笋片、咸菜，焖至汤浓鱼熟，加盐和味精少许，出锅即成。此菜主料咸菜和黄鱼均以鲜见长，加之烹调得法，使两鲜合一，鱼肉鲜嫩，咸菜清香，汤汁稠浓，味道十分鲜美。

在浙江温州一带，有许多冠以敲字的鱼味菜。温州人逢年过节，总要用特制的小木槌敲几张"敲鱼"，烧一味敲鱼菜，其中又以三丝敲鱼最为出名。关于三丝敲鱼的来历，还有一个美妙的民间传说。相传很久以前，温州某寺方丈在海上遇难。小和尚悲痛万分，每日手持木鱼到海边诵经悼念师父。一天，他在海边发现一大群黄鱼，便想到师父或许是被黄鱼

soup with pickles. The local people call pickles as pickled chicken. And there is a story telling its origin. The legend said during the late 1390s in the Ming Dynasty, Wang Jin, the newly appointed magistrate of Mingzhou (now Ningbo city), was an honest and upright official. He often invited local officials and wealthy people to his home and treated them with porridge and pickles. During the meal, he also chanted a poem titled Ode to Chicken Pickled : "Once I think of people's hardships, the taste of the pickles is like that of chicken." Since then, the pickled chicken feast has grown in popularity and pickles have been hailed as "pickled chicken." In fact, the pickles in this dish is from potherb mustard, an important ingredient of many home dishes in Ningbo. There is a local saying, "No one could live well without pickles for three days." To cook this dish, first scale and wash a fresh big yellow croaker, then heat some lard in the wok. Fry the yellow croaker, add Shaoxing wine, scallions, ginger, fresh soup, bamboo shoots, pickles, then allow them to simmer until the soup gets thick. Finally, add some salt and MSG and the dish is ready. The dish combines pickles with yellow croaker and it not only keeps the freshness and tenderness of the fish, but also adds more savor to pickles and creates a tasty soup.

In Wenzhou City, Zhejiang Province, when you look through the menus in local restaurants, you will find that there are many fish dishes that contain the word "knock. " When Wenzhou people celebrate their holidays, they often use a special small mallet to knock a few fishes to cook special dishes including the best known knocked fish with three slices. There is also a wonderful folklore about the origin of the dish. A long time ago, a temple abbot from

吃掉，于是脱下袈裟，网起黄鱼，剖皮去骨，放在木鱼上敲打，以此为师父报仇。他把敲过的鱼丢在海滩上晒。四十九天后超度期满，海滩上晒满了敲鱼干。离开时，他带走几片做纪念，其余的就分给当地居民食用。人们把敲鱼干拿去煮汤，觉得滋味很鲜美，就仿做起来。后来有些厨师在烹敲鱼干时，加入了瘦肉、鸡脯、火腿丝，于是就有了"三丝敲鱼"这道名菜。

在制作这道菜时需先将新鲜的黄鱼剖肚，去骨刺，然后以木棒敲成饼状薄片，置于阳光下曝晒成干。食用时，将敲鱼片切，和青菜一起入沸水氽熟，后沥干。锅中加清汤，放进鱼片、青菜心、精盐、料酒，用中火烧沸，撇去浮沫，再配以瘦肉丝、鸡脯丝、火腿丝，加调味品，最后淋上熟鸡油，起锅入盘。其成品肉质嫩滑，色白汤清，香醇可口。

除了鱼虾贝类，浙菜中以海蟹为主料的菜品也十分繁多，比如颇负盛名的红膏呛蟹。据说这道名肴还是先人意外所获。从前，渔民出海捕蟹，船上没有冰块用来保鲜，于是便想出把蟹倒进船舱，灌入海水，再加一点盐腌制的办法。这种用盐将蟹活活"呛死"的方法却取得了意想不到的效果：人们发现之后捞起来吃时，蟹的口感居然很好。于是这门手艺便从船上传到了岸上，由厨师加以改造，渐渐成为了一道名菜。

Wenzhou was killed at sea. A priestling was in great grief and went to the beach every day with a wooden knocker to the beach to mourn his master. One day, he found a large group of yellow croakers at the seaside. He thought that the master might have been eaten by croakers, so he took off his kasaya❶ to net the fish. After he succeeded, he peeled them, removed their bones, and beat them with his wooden knocker to revenge for his master. He then threw the knocked fish on the beach and under the sun. After forty-nine days, and the beach was covered with dried fishes. When he left, he took a few to commemorate, while the rest was distributed to local residents. Some used the dried fish in cooking soup and found it so delicious. So, this cooking method became popular. Later, some chefs added sliced lean meat, chicken breast, and ham when cooking dried fish, and that's how the famous dish, knocked fish with three slices, came into being.

To cook the dried fish, first, wash a fresh yellow croaker, remove its internals and bones, and then use a wooden stick to beat it into a thin slice. Dry it under the sun. Then, wash and cut the dried fish into fillets, blanch them in boiling water with some Chinese cabbages, and then remove and drain. Next, add some clear soup in the wok, put in the fillets, cabbage heart, salt, cooking wine, and boil it in medium heat. After it boils, remove the froth, and then put in the sliced lean meat, chicken breast and ham as you add the seasonings. Finally, finish the delicacy with a touch of cooked chicken oil. The tender and smooth taste of the meat and the clear soup color make the dish one of the favorites of Wenzhou people.

In addition to fishes, shrimps and shellfish, Zhejiang Cuisine

❶

Kasaya（袈裟）, a patchwork outer vestment worn by a Buddhist monk.

also includes a wide variety of dishes with sea crabs as the main ingredient including the famous choked portunid. It is said that this dish was invented inadvertently. Once upon a time, fishermen went to the sea to fish for crabs. Since there were no ice cubes on the boat to preserve the freshness of the crabs, they came up with the idea of pouring the crabs into the cabin with some seawater and adding a little salt to choke them to death. However, this method of using salt to kill crabs had an unexpected effect: people found that the crabs had a delicious taste after they were choked. Therefore, the cooking method was passed to the shore and was improved upon by chefs. Gradually, it became a famous dish. To cook this dish, first, wash the portunid with cold water, then add prepared marinade and mix well. Put them in a sealed container to choke and marinate. After that, take one crab out from the container, and crush the crab claws with the back of a kitchen knife, and remove the gills, the stomach and the carapace under its belly. Next, cut it into pieces according to the pattern of its claws. A good choked portunid is neither salty nor fishy, and each piece of crab has reddish roe and a delicious flavor.

The literati consistently depicted the world through literary works based on their unique vision. Given that Zhejiang has gathered a significant number of famous literati through out its history, there is a considerable volume of books on Zhejiang Cuisine. An example is Yuan Mei's *Recipes in Sui Garden*. Born in Hangzhou, Zhejiang Province, the famous poet and food critic in the Qing Dynasty was very fond of tasty food. He wrote poems and essays for pleasure and enjoyed his companionship with friends.

制作时需先将梭子蟹以冷水洗净，加入配制好的呛料密封腌制，待到吃时捞出呛蟹，用刀背将蟹钳压碎，去脐、腮、胃，按蟹爪的纹路，切作一股一股。一盘好的红膏呛蟹要求不腥不咸，且每块蟹肉都要带有红膏，色泽亮丽，口味鲜香，做到真正的色香味俱全。

文人执笔，落字生花，关于浙菜，留下的书文也不在少数，一如袁枚的《随园食单》。袁枚是杭州人，富盛名而好美食。他以诗文自娱并广交文友，一生著作颇丰，《随园食单》是他不经意间所作书籍，却是中国饮食理论的杰出代表，在中国饮食文化史上有着里程碑的意义。他在书中对浙菜的记载颇多，比如鲅鱼豆腐、台鲞煨肉、宁波的海蜇、江瑶柱，乐清、奉化的牡蛎等。

浙菜当中的鳗鱼鲞、黄鱼羹、敲鱼菜、呛蟹膏，有故事也有美味。在世界范围内，同样的美食与故事也有迹可循。日本有江户时代完善起来的食鳗鱼传统，日本的鳗鱼饭至今仍名声远播，老店林立。泰国有着源自吞武里王朝时口味独特的冬阴功汤，被奉为国汤。挪威有传承自维京时代的制鱼干技艺，今天当地居民仍会用干净的布裹好干鳕鱼，用铁锤砸碎，剔除鱼骨、鱼皮，享用鱼肉。在法国，食生蚝从皇室传入民间，路易十四对贝隆生蚝牛奶口感的嗜好，恰似浙江

Yuan wrote many books during his lifetime including *Recipes in Sui Garden*, which he wrote by accident. The book has become an outstanding representation of Chinese diet theory as well as a milestone in the history of Chinese food culture. In this book, he engages in a detailed discussion of Zhejiang Cuisine and he talks about dishes including stewed abalone with tofu, stewed dried fish with pork of Taizhou, Haiyan[1] and dried scallops of Ningbo, oysters of Yueqing and Fenghua [2] among others.

Dried eel fish, yellow croaker soup, knocked fish, and choked portunid and almost all other dishes in Zhejiang Cuisine have stories accompanying their delicious tastes. It is also noteworthy that similar food and stories exist around the world. Japan has a tradition of eating eels that was perfected in the Edo period. Till present day, the Japanese eel rice is still well-known at home and abroad, and old shops selling eel rice are everywhere. Thailand has a unique soup called Tom Yum Kung that can date back to the Thonburi period and it is regarded as the national soup. The Norwegians inherited the traditional skill of making dried fish right from the Viking era. Today, local residents still wrap dried cods with clean cloth, smash it with an iron hammer, remove the fish bones and skin before enjoying them. In France, eating fresh oysters was introduced from the royal family to the folks. Louis XIV, the Sun King, loved the milky taste of Bélon so much. His love can be likened to Zhejiang people's preference for choked portunid.

[1]

Haiyan (海蜒), a small fish produced in Ningbo, Zhejiang Province, tastes like shrimps.

[2]

Taizhou (台州), Ningbo(宁波), Leqing (乐清), Fenghua (奉化) all belong to Zhejiang Province. And Fenghua is a district of Ningbo City.

人对红膏呛蟹的偏爱。经纬不同，地域不同，但人们对于海味和其中隐含文化的认同是一致的。

　　钻木取火以烹煮，结罟养牲以庖厨，饮食是人类发展进步中关键的一环，文化亦然。由饮食至文化，本身就是人类的一大发展进步。浙菜中所包含的神话、史话、趣闻、书文等，有内蕴，也有情致，这种饮食与文化结合而生的气息，正是千里江山图中的一抹神韵。

Despite the different latitudes and longitudes, people all appreciate the deliciousness of seafood and the implicit cultures.

Our ancients drilled wood to make fire and they fished and raised animals to provide raw materials. Diet is indeed a key part of human development as well as culture. Going from diet to culture has been a major progress for mankind. The myths, historical stories, anecdotes, and literary works contained in Zhejiang Cuisine have both implication and beauty. The combination of diet and culture is a touch of charm in the thousands of miles of rivers and mountains that span through the entire China.

醇美，八闽的调和

Chapter

Six

Fujian Cuisine:
Mellow Harmony

　　福建，古为闽地，宋时，有六州两军，元时分八路，明改为八府，故又称八闽。福建全省多山，又面朝大海，自古以来，当地人就习惯于依山傍水而居，饭稻羹鱼，与海洋结下了深深的情愫。

　　依山傍海带来的是丰富的食材，这是闽菜发展的基础。顺着历史脉络看，晋时中原战乱频繁，"永嘉之乱"后，大批中原士族南渡，始入闽地，并将中原的科技文化带入，促进了当地的发展。至晚唐时，河南光州固始的王审知兄弟带兵入闽建立"闽国"，对福建饮食文化的进一步发展繁荣产生了积极的影响。明清以后，海上丝绸之路带来了更多元的文化，也给福建饮食注入了新鲜血液，终成就特色闽菜。闽菜早期就是福州菜，后发展成福州、闽南、闽西三种流派。闽菜清鲜和醇、荤香不腻而又汤路广泛，是为醇美，在烹饪界中独占一席。同时，相比八大菜系中的其他菜系，闽菜又

Fujian is also known historically as Min due to the "seven Min tribes" that inhabited the area during the Zhou Dynasty. It was, however, during the Song Dynasty, divided into six small prefectures and two military prefectures and then changed to eight prefectures in the Yuan and Ming dynasties. That is why it is called "Eight Min." Since ancient times, people in Fujian have been engaged in rice farming and fishing as they live near the sea and mountains. This means they have been deeply attached to the sea.

Thanks to its geographical advantage, Fujian is rich in ingredients that constitute a solid foundation for the development of Fujian Cuisine (Min Cuisine). Historically, the first wave of the noble class moving to Fujian happen in the middle Jin era during which wars broke out frequently in Central China, especially the "Yongjia Rebellion." This southbound move brought Central China's culture to the Min region, greatly driving the local development. During the later years of the Tang Dynasty, there were several political upheavals. Wang Shenzhi and his brother, who were from Gushi county, Henan Province (present day Xinyang city in Henan Province), led their troops to the Min region to set up an independent Kingdom of Min with Fuzhou as its capital. This action promoted a second wave of Fujian development in food culture. After the Ming and Qing dynasties, the maritime silk road not only brought diverse cultures, but also injected fresh blood into the food culture, eventually forming a unique cuisine—Fujian Cuisine. Though the cuisine originally referred uniquely to Fuzhou dishes, it has gradually developed into three branches: Fuzhou, South Fujian, and West Fujian dishes. Rather than the blind

另有三大特色，其一长于红糟调味，其二善于制汤，其三擅使糖醋，是为调和。

闽菜的红糟调味实际上源自中原。在唐代以前中原地区已开始使用红曲作为烹饪的作料。红曲由中原移民带入福建后，红糟也成了闽菜常用的作料。所谓红糟，即酿制红曲酒时经发酵产生并筛滤得到的酒糟，是闽菜独有的调味方法。红糟的烹调方法多样，有炝糟、淡糟、拉糟、醉糟、炝糟、灯糟等十多种，采用此法所烹制的佳肴也颇多，比如淡糟鲜竹蛏和炝糟菊花螺。淡糟鲜竹蛏以福建当地所产竹蛏为主料，烹饪时先将鲜竹蛏肉洗净切片，与冬笋片、香菇片入沸水略氽捞出，加入黄酒腌制后滗去酒汁。热锅下油，下姜蒜稍煸，再加红糟煸炒，烹入黄酒、精盐、味精、白糖、上汤，加湿淀粉勾芡，再倒入竹蛏、冬笋和香菇，翻炒装盘即成。其成菜色泽淡红，蛏肉丰腴脆嫩，甘鲜怡人。

福州有句俗语："要食海味，黄螺、珠蚶"，可见黄螺在福州人心目中的地位。炝糟菊花螺便以黄螺为主料，先将黄螺肉改十字花刀，焯水待用，青瓜切块，焯水装盘。热锅下油，入姜末炒出香味，下红糟煸炒至熟，加入黄酒、虾油、白糖、味精，炒至糟黏稠后倒入黄螺，颠炒几下，淋上芝麻油，装盘即成。烹制此菜时，一是黄螺需改十字花刀，使其

pursuit of taste stimulation, Fujian Cuisine always seeks to produce a mellow and plain flavor and diversity in soups. This way, it can win itself a place in the Eight Major Cuisines of China. At the same time, compared with other seven cuisines, Min Cuisine has three notable characteristics: adept in using red yeast rice for flavoring, good at making soups, and skillful in blending sugar and vinegar to balance the flavor.

The use of red yeast rice is originated from Central China where the practice of using it as a seasoning started before the Tang Dynasty. Immigrants took it with them and then it was commonly used as a unique material for flavoring in Fujian Cuisine. The red yeast rice refers to the wine lees from fermented red glutinous rice wine. The cooking methods are varied and they include slight pickling, quick frying and deep pickling, while the dishes derived from them are countless, including fresh razor clam in rice wine and chrysanthemum-shaped babylonia lutosa in rice wine. The main ingredient of the fresh razor clam in rice wine dish is local fresh razor clam. To prepare this dish, clean fresh razor clam, then slice it. Boil sliced winter bamboo shoot and mushrooms together for a while. After that, add Shaoxing wine to pickle them well, add ginger, garlic, and red yeast rice to sauté together with oil. Next, add Shaoxing wine, salt, MSG, sugar, clear soup, wet starch, and stir well. Finally, mix razor clam, winter bamboo shoots and mushrooms to sauté quickly and serve. The razor clam meat is savory and tender and has a light red color. It is delicious and the taste will linger in your mouth.

A popular Fuzhou saying "Babylonia lutosa and mud clam

易于成熟，且形似菊花，二是需旺火快炒，保持肉质脆嫩，糟香浓郁。

长期以来，闽厨把烹饪和对原料质鲜、味纯、滋补的追求联系起来，总结出汤菜为上的结论，故闽菜素来以汤菜见长。闽菜的传统菜谱是：两真果、四冷碟、一大拼、六炒盘、八大件、两点心、一甜汤。其中八大件是主菜，且都以汤菜的形式出现。说起汤菜，不得不提到一个故事。相传一百多年前，聚春园的创始人郑春发原是福建布政使周莲的家厨，在一次官宴上学到一道菜的制作方法，后几经调改，终于研发出名菜配方。一日，几位秀才到聚春园吟诗作赋，郑春发送上此菜。坛盖揭开满堂荤香，秀才们无不拍手称奇，吟诗道："坛启荤香飘四邻，佛闻弃禅跳墙来。"此菜因此得名"佛跳墙"。

佛跳墙曾作为国宴名菜招待过许多中外嘉宾，是一道誉满中外的福州传统名菜。在制作时需选用上等海参、鲍鱼、干贝、鱼翅、鸡肫、猪蹄筋、火腿、鸭肉、羊肘、鸽蛋等多种考究的原料，佐以花菇、冬笋、绍酒、冰糖、桂皮、姜片、白萝卜等辅料。烹制时先将几十种主辅料经过加工处理后，分批分层放入绍兴酒罐中，用荷叶盖口封密，再盖上一个小碗，然后把酒罐放在木炭炉上，用文火慢慢煨制数小时即成。

must be your first choice in seafood" illustrates the genuine love of the people of Fuzhou for babylonia lutosa. The main ingredient of the dish chrysanthemum-shaped babylonia lutosa in rice wine is, of course, babylonia lutosa. To cook the dish, first, cut the babylonia lutosa crosswise, blanch it then set it aside. Next, cube the green cucumber and blanch. Next, put chopped ginger into a heated wok and sauté with oil until it gives out a savory smell. Thereafter, add red yeast rice wine lees to stir fry until it is well done. After that, mix Shaoxing wine, shrimp oil, sugar, MSG, and sauté until the red sauce thickens. Then pour babylonia lutosa to sauté quickly and stir well. Finish by drizzling sesame oil into it. During the process, chefs can never forget to cut the babylonia lutosa crosswise like Chrysanthemum to make it easy to cook. And they cannot forget to stir-fry over high heat to keep the meat crisp and tender and to retain the rich sweet taste of red yeast wine lees.

Given their pursuit of fresh, pure, and nutrition-rich raw materials, Fujian chefs found that nothing apart from soups can best keep the original flavor and nutritive value. As a result, they advocated for soups. The traditional recipe for Fujian Cuisine comprises two main kinds of fresh fruits, four plates of cold dishes, one big plate of assorted dish, six plates of fried dishes, eight covered dishes, two kinds of desserts and a bowl of sweet soup, of which eight covered dishes are main courses in the form of soups. There have been a great many stories about soup dishes, of which the most famous being the story of Buddha Jumps Over the Wall. It is said that more than one hundred years ago, Zheng Chunfa, who later became the originator of Ju Chunyuan (a famous

hotel in China), was formerly just a private chef of Zhou Lian, a commissioner of Fuzhou city. Zheng Chunfa once learned a special recipe during an official feast held in Zhou's mansion, after which he made significant efforts to perfect it. One day, when several scholars gathered in Ju Chunyuan to talk about poems, Zheng took the opportunity to present the dish. It filled the entire garden with a savory smell, which made it impossible for the scholars to do nothing apart from singing their praises in a poem as follows: " The delicious dish is so tempting that even a Buddha will give up his sutra and jump over the wall for a taste." The dish, therefore, got its name as Buddha Jumps over the Wall.

As a representative of traditional Fujian Cuisine, Buddha Jumps over the Wall has made its way onto the state banquet menu and its reputation has increased as it has been served to guests at home and abroad. To cook this dish, only top-grade materials are allowed and they include sea cucumber, abalone, dried scallop, chicken gizzard, shark's fin, pork tendon, ham, duck, sheep cubit, and pigeon egg. Other ingredients include flower mushrooms, winter bamboo shoots, Shaoxing wine, rock candy, cinnamon, sliced ginger, and moolis. First, process all the ingredients well then put them slowly into a Shaoxing wine jar layer by layer. Seal well with a lotus leaf. After that, cover it with another small bowl. Finally, put the jar over a charcoal burner and simmer it over low heat for a few hours. Then Buddha Jumps Over the Wall is ready to serve. Nothing but a Shaoxing wine jar can help perfectly integrate all of these ingredients and preserve the smell during the cooking process. Thus, when opening the jar, the savory smell assails diner'

此菜集几十种原辅料于一罐，制作时无气味泄露，揭盖时浓香扑鼻，入口时质厚不腻，既调和各味之精华，又保持各自的特色，堪称闽菜醇美的代表。

除了"闽菜之首"佛跳墙，被誉为"闽菜之后"的鸡汤汆海蚌也是一道著名的汤菜。此菜中的"鸡汤"并不是单纯的"鸡"汤，而是以老母鸡、鲜牛肉和猪里脊肉为原料，加上诸多工序煲出的特制"三茸汤"。"海蚌"则特指福州长乐漳港所产的海蚌。制作时先将蚌肉切成薄片，入滚水汆至五成熟取出，加黄酒稍加腌渍后沥干酒汁，再以鸡汤稍浸并滗净汤汁。蚌肉分盛于小碗中，上席时，鸡汤烧沸，加精盐调味，淋在蚌肉上即成。此菜讲究现淋现吃，鸡汤清澈见底，蚌肉脆嫩鲜美，味道极佳，余味悠长。

闽菜中很多菜肴会用到糖和醋，在力求保持食材原汁原味的基础上，用糖以去腥膻，用醋以增爽口，因而甜而不腻，淡而不薄。把握两者的均衡本就不易，合两者为一，创出新口味也不易，用于海鲜则更不易，比如糖醋萝卜蜇。这道凉菜经常以开胃菜的形式出现在闽式宴席上，制作时先将白萝卜、海蜇皮和辣椒分别切丝，将海蜇皮丝以清水泡淡，白萝卜丝和辣椒丝放入另一个清水盆，加入精盐浸泡二十分钟。将泡淡的海蜇皮丝捞出，和萝卜丝、辣椒丝放在同一盆里，

noses. The dish is plump but not greasy and it blends various flavors in perfect harmony albeit preserving their own characteristics. Possessing such a mellow and plain taste, it is no wonder this dish is a representation of Fujian Cuisine.

Locals often consider the dish Buddha Jumps Over the Wall the king of Fujian Cuisine, the queen, of course, being another famous soup dish—chicken soup quick-boiled with sea clams. The "chicken soup" here means a specially-made "assorted soup, "whose main materials comprise of an old hen, fresh beef and pork tenderloin. The ingredients are well stewed in a variety of cooking processes. Meanwhile, "sea clam" here refers, in particular, to the clam produced in Zhanggang port, Chang'le district, Fuzhou city. To cook the dish, slice sea clam and quickly boil to medium rare. Then marinate the clam slices in Shaoxing wine for a little while and slightly soak them with chicken soup before draining. Next put them in separate bowls as you boil the chicken soup and add some salt. Finish by sprinkling the hot chicken soup over the calm slices in order to make the dish distinctive. Clear chicken soup goes well with crisp and tender clam and it leaves a long-lasting flavor in your mouth.

While remaining the natural flavor, many dishes of Fujian Cuisine often use sugar to remove the fishery smell and vinegar to increase the tastiness. So, the dish is neither greasy nor tasteless. Rather, it is sweet and plain in a way that suits everyone's taste. It is not easy to strike a balance between the original flavor and seasonings, let along to create a new dish with seafood. Fujian

搅匀捞起。用清水多次冲洗后，挤干水分，加香醋、白糖拌匀，待白糖溶化，装盘即成。此菜刀工细腻，色彩丰富，酸甜爽口，具有很好的开胃功效。

此外，肉米烂鱿鱼也是一道典型的糖醋海鲜菜。烹制时先将鱿鱼切块，入沸水略氽，捞起沥干，然后热锅下油，下香菇、冬笋、西红柿、五花肉米、葱米煸炒，冲入骨汤，下酱油、白糖、味精，煮沸后以湿淀粉勾芡，滴入香醋，放入鱿鱼片搅匀，最后装碗，撒上胡椒粉、淋上芝麻油即可。其成品色泽淡红，酸甜带辣，软滑爽脆，十分可口。

宴席上的佳肴固然美味，然街边的小吃也别有风味。闽南沿海有一种著名的小吃，叫做土笋冻，以海边一种状如蚯蚓的软体动物土笋为主料，将其洗净后煮熟，分装小酒盅中，冷凝后倒入盘子里，形成胶状，加蒜泥、沙茶、辣酱、香醋等蘸食，其味道鲜美，风味特殊，食者无不赞誉。产于海边泥沙中的黄泥螺也是小食摊常售之物，其烹制简单，或煮或炒，以蒜泥、香醋、酱油、辣椒酱佐之即可，食者用手指夹起小螺，对螺嘴吸吮，只听"啾啾"的响声，螺肉入口，十分鲜美细嫩。

另有两种与抗倭有关的小吃，那就是蛎饼和鼎边糊。蛎饼是闽中、闽北群众喜爱的早餐小吃，需用大米、黄豆加水

Cuisine however, is undoubtedly an "expert" in this respect. A typical example is the dish sweet and sour jellyfish with mooli, which often appears in the Fujian-style banquet as an appetizer. To cook the dish, first shred mooli, jellyfish, and chili. Then soak the jellyfish in clear water. After that, put the mooli and chili into clear water and add salt and soak for twenty minutes. After that, put the three together and rinse thoroughly with clear water before drying. Finish by mixing them well with vinegar and sugar. If your cutting skills are good enough, the dish will be colorful in appearance, refreshing in taste and extremely appetizing.

Sweet and sour squid with mashed streaky pork is another exemplary dish. To cook the dish, first, cube the squid, boil and quickly drain them. Next, add mushrooms, winter bamboo shoots, tomatoes, mashed streaky pork and chopped scallions in a wok heated with oil and stir-fry. Thereafter, pour bone soup and add soy sauce, sugar, MSG and boil the mix. Then mix wet starch until the sauce thickens, add vinegar and squid and stir well. Finally serve with pepper and sesame oil. The dish which has a light-red color tastes sweet, sour, and a little spicy and thus it suits the taste of people from different regions.

Besides delicacies in the feast, street snacks that embellish Fujian Cuisine equally associate us with tastiness. For example, in the southern coastal area of Fujian, there is a famous snack named Sea Worm Jelly. The main ingredient is the peanut worm, a type of mollusk that has the same appearance of an earthworm. To cook the dish, first wash clean peanut worms and cook them thoroughly. Then put each worm into a wine cup and condensate

磨成浆，加入精盐少许，搅拌均匀，将一汤匙的浆倒入小铁勺中，内放新鲜的海蛎和葱花，然后再倒上一汤匙的浆盖在海蛎上入油锅炸制，至蛎饼胀起，两面均成金黄色时就可捞起。其色泽金黄，皮酥味美，深受当地人的喜爱。

又因福建人爱喝汤水，因此蛎饼也不单吃，最常配现做现吃的"鼎边糊"。明朝嘉靖年间，福州沿海一带常遭倭寇骚扰，戚继光带兵入闽剿倭寇，受到当地民众的拥戴。据说有一天，戚家军到了福州南郊，当地乡民摆下八仙桌，主动送来大米、鱼肉、香菇、虾皮等，准备招待众将士。就在此时，又有倭寇来袭，队伍马上准备出战。老百姓却无论如何要让将士们吃饱了再去打仗。不知谁灵机一动，在烧着菜的锅边烙起米浆来，不消一刻钟，一锅又一锅的鼎边糊就出来了。众将士吃饱后奋勇上阵，大胜而归。此后这种吃法流传开来，用蚬子汁为汤，在锅里烧开取其鲜味，再把磨好的米浆沿着锅边一圈浇过去，米浆在锅边烫成干皮后用锅铲刮到汤里，加入芹菜、葱、虾皮、香菇等，烧开后起锅就是一盆滚烫的鼎边糊了。

福建属亚热带季风气候，舒适宜人，降雨量又较为充沛，因此水产颇丰。相同的环境往往会造就类似的饮食，同属这一气候的国外城市也有着同闽菜相类似的菜肴。比如在韩国

into colloid before serving. Finally serve with mashed garlic, barbeque sauce, thick chilli sauce, vinegar, etc. Thanks to its savory smell and delightful flavor, the dish gets rave reviews among diners. Another savory snack is made easily by using locally produced Korean mud snail that does not cost much. Whether it is sautéed or boiled, the snail meat is tender and tasty when simply served with mashed garlic, vinegar, soy sauce, or chili sauce. All that diners need to do is to take up the snail with their fingers and suck the meat with "chirp" sound.

Two other Fujian snacks deep-fried pancake with oyster and rice paste in shrimp soup, were closely related to battles against the Wokou ❶. Deep-fried pancake with oyster is such a popular breakfast snack that is especially loved by people in the central and northern Fujian. To make this snack, grind rice and soybean mixed with water into a thick liquid. Then add a little salt and stir well. Next, scoop one tablespoon of thick liquid and combine with fresh oyster and scallions. After that, cover with another scoop then fry it in the hot oil until it turns golden brown on both sides. The bright color and crispy taste of the pancake win the locals' favor.

Meanwhile, locals normally serve the pancake with another freshly-made soup dish rice paste in shrimp soup to soothe the dry taste. There is a story about the soup. During the period from AD1522 to AD1566, under the reign of Emperor Jiajing of the Ming Dynasty, Qi Jiguang (a military general) led the defense against Wokou on the coastal region of Fujian where pirate activities were

❶

Wokou (倭寇) were pirates who raided the coastlines of China and Korea from the 4th century to the 16th century. They came from Japanese, Korean, and Chinese ethnicities which varied over time and raided the mainland from islands in the Sea of Japan and East China Sea. Wokou activity continued in the Ming Dynasty of China and peaked during the mid-1500s.

frequent. Qi and his troops were so welcomed that local people often volunteered to send rice, fish, mushrooms, and shrimps to them as a treat. One day, Qi and his troops went to the southern suburbs of Fuzhou city， and no sooner had they arrived than the locals got ready to cook for soldiers. Suddenly the Wokou attacked and the troops decided to get into battle immediately. The locals nevertheless would not let them go with an empty stomach. So someone got the inspiration to quickly pour rice milk on the edge of the boiler, and within fifteen minutes, baked it to rice paste. Fighters fought bravely against the enemy after eating their fill and finally won the battle. Thereafter, the dish spread widely ever since. To cook this meal, boil clam soup and then pour rice milk into the wok along the edge and bake them until they turn crispy. Shovel the rice crisp into the soup . Finally, add celeries, scallions, dried small shrimps, mushrooms, etc. and boil. The hot and savory rice paste in shrimp soup is finished.

With a subtropical climate and plentiful rain, Fujian is quite livable and has an abundance of sea products. Given that dishes born in a similar environment and climate share the same characteristics, Fujian Cuisine definitely has its friends overseas. For example, in Busan, Korea, the flour, seafood and scallions combination has been very popular among both locals and tourists in Dongrae Hot Spring. People in New Orleans opt to cover fresh oyster with flour and fry until it turns golden brown and becomes similar to the deep-fried pancake with oyster in Fujian. While in Nassau, the capital of Bahamas, people eat well-cooked whelk meat, with lime juice and seasonings served as a dipping sauce， which has

釜山，面粉加海鲜和葱煎成的海鲜葱煎饼，是东莱温泉的著名小吃。在新奥尔良，人们喜欢将生蚝裹上面粉，炸至金黄食用，这和福建的蛎饼十分相似。在巴哈马的首都拿骚，以熟海螺肉蘸酸橙汁、调味料食用，为的是去除腥味，享受鲜美。在巴西，炖鱼则是受到追捧的美食，以当地海鱼配番茄、洋葱在陶煲中慢火炖制而成，而在入煲之前，用酸橙汁腌制必不可少。

如今，海上丝绸之路重新开通，闽菜也将顺着这条路走向更为广阔的世界，无论是餐桌宴席，还是街边小吃，闽菜都会继续以其独特的烹饪技法，调和不同食材的特色，释放出最醇美的味道。

the magic of keeping it savory and free from the fishery smell. What distinguishes people in Brazil is their obsession for roast fish. After marinating with lime juice, local marine fish, together with tomato and onion, is putting into a casserole and stew slowly over low heat.

Nowadays, with the Maritime Silk Road re-opening up, Fujian Cuisine, together with the rich culture it embodies will be introduced into a wider world. It will continue to harmonize different flavors of ingredients with the help of its unique cooking skills. The dishes of Fujian Cuisine, either exquisite banquet or popular snacks, provide diners with still more mellow and plain tastes.

精细，南粤的丰盛

第七章

Chapter

Seven

Guangdong Cuisine:
Refined Abundance

先秦古籍中，对长江以南沿海一带的部落统称为"越"，广东被称为"南越"，因通假和区分的缘故，后改为"南粤"，时至今日，广东仍简称"粤"。因此广东菜就有了一个响亮的名字——粤菜。

《淮南子》有载："越人得髯蛇，以为上肴，中国得而弃之无用。"广东人的饮食，食材之丰富，由来已久，而这注定了粤菜的丰盛。粤菜选用的主料广博奇杂，选用的配料纷繁多样。飞禽走兽、山珍海味皆可入菜，但获取食材易，制成佳肴难。因为从广博的主料中选择合适的品种，从众多的配料中选择独特的味道，再辅以恰到好处的装饰造型，无一不需要厨师的比较、把控、选择、搭配，而这恰是粤菜的精细。菜肴众多的粤菜中，海味至关重要，我们就从这里讲起。

早晨，闹市间，人们已纷纷迈出家门，向茶楼奔去。

According to the Pre-Qin classics, there was a mixture of tribal groups known as "Yue" (越) on the south coastal areas of Yangtze River, of which Guangdong was called "South Yue." The name was later changed to "South Yue" (粤, the same pronunciation as 越) in order to differentiate it from other tribes. Even to this day, Guangdong Province has kept its short name as "Yue" (粤), a name which provides Guangdong Cuisine with a recognizable name— Yue Cuisine (Cantonese Cuisine).

According to the classic *Huai-Nan-Zi*, "When people in the State of Yue catch a big snake, they use it to make a delicious dish. However, a big snake has no use in Central China." These words testify the fact that Guangdong is home to countless cooking ingredients for a long time and Guangdong Cuisine is, of course, full of hearty dishes. Cantonese are known to have an adventurous palate, and the selection of ingredients is therefore so varied that everything that walks, crawls, flies, or swims, can be made into a dish. Even if ingredients are available, dishes are hard to prepare because it is very challenging for chefs to skillfully select matching ingredients to deliver the best taste and to present a delicate plate. It is chefs' crafts comparison, selection, and collocation that lead to the refinement of Guangdong Cuisine. This refinement of Guangdong dishes also includes an important component—seafood. Thus, we will start here.

At dawn, Guangdong people leisurely leave their homes and, one after another, head for a tea house. Morning tea is, of course, Cantonese' favorite, but it is not only about tea. Before starting work, urbanites, for instance, prefer to chat with each other and

广东人最喜吃早茶，但又不只是吃茶这么简单，一盅两件，既是一种生活方式，又是一种社交方式，更是一种独特的饮食文化。早茶的茶点中，虾饺与肠粉是最为常见的食物之一。虾饺原本出自伍村五凤乡，后城内茶馆将其引进，加以改良，以明虾仁为主料，讲究的是面皮晶莹透亮，口感柔韧鲜美。

关于肠粉，还有一则故事。相传乾隆下江南时，经不住大臣纪晓岚的蛊惑，专门转到罗定州吃龙龛糍。当吃到这种嫩滑的龙龛糍时，乾隆赞不绝口，并乘兴说："这糍并不算是糍粑，反而有点像猪肠子，不如就叫肠粉吧。"后来他回到了京城，还对罗定的肠粉念念不忘，而肠粉也因此在广东传开了。传闻真假已不可知，但肠粉最初的确是被称为"龙龛糍"，且在广东境内就有诸多分类，其中普宁的潮汕肠粉最能显出丰盛，常见的搭配材料有生蚝、香菇、白萝卜干、干鱿鱼、鸡蛋、生菜、西洋菜、空心菜、肉沫、鲍鱼等。肠粉粉质的嫩滑与海鲜、蔬菜、肉类搭配在一起，美味而鲜香。

荔枝湾，入夜时分，灵活的花艇靠向岸边，开始向归途中的人们出售新鲜出锅的米粥，而这正是旧时广州常见的场景。彼时花艇上的热粥，如今有了正式的名字——艇仔粥。

enjoy life a little bit over one pot of hot tea and two plates of snacks. Morning tea has become not only a lifestyle and a social way of life, but also a unique food culture. Among all the refreshments, Steamed Prawn Dumplings (Har Gow) and Steamed Vermicelli Roll seem to be the most common. Steamed Prawn Dumplings first appeared in Wu Feng Village near Guangzhou outskirts and was later introduced and improved by tea houses in the city where dim sum chefs worked. With fresh prawns as the main ingredient, chefs finally made the dish an exquisitely delicious snack with a transparent and smooth wrapper as well as a pliable and savory taste.

As for Steamed Vermicelli Roll, there is another story. It is said that when Emperor Qian Long (AD1737–1796) went to Jiangnan (the south lower reaches of the Yangtze River), he was encouraged by Ji Xiaolan, his minister, to make a detour specially to Luoding (Luoding city in Guangdong Province) to enjoy longhe glutinous rice cake, a local delicacy. Soaking himself in the soft and tender taste, the Emperor was full of praise, so he said in spirits: "It looks more like chitterlings rather than glutinous rice cake, why not call it Steamed Vermicelli Roll. " The enticing aroma lingered long in the emperor's mind even after he went back to the palace. After that, the dish spread widely in Guangdong Province. Whether the story is true or not, Steamed Vermicelli Roll was initially known as longhe glutinous rice cake and there are numerous types in Guangdong. They include Chaoshan style Steamed Vermicelli Roll, which is common in Puning city and also the most sumptuous type with ingredients including oysters, mushrooms, white dried

原本，正宗艇仔粥，是水上人家用新鲜打捞的鱼虾蟹蚬螺等，杂七杂八，汇入一煲粥里熬出来的，且最初的艇仔粥每天用料不尽相同，渔民捕获什么就加入什么，既新鲜又有惊喜，粥水鲜甜无比。今天，艇仔粥已有了固定的做法，用粳米熬好底粥，要出锅时，加入新鲜的虾或鱼片、炒花生、海蜇、鱿鱼、浮皮、葱花、姜等，用底粥煮沸，加入调料，盛出即食。河鲜与海鲜的搭配，带来了更有层次的鲜味，让人回味无穷。

湿热的气候使广东人钟爱煲汤与喝汤，这是他们传承了千年的养生秘法，也因此老火汤已经成为当地人生活中不可缺少的部分。热爱所带来的力量是巨大的，老火汤单是有记载的汤谱就达数百种，似乎万物皆可入汤，其丰盛程度可见一斑。与此同时，煲汤必用砂锅，火候讲究变化，关火后要焖够时长，不同的汤搭配不同的中药等诸多讲究，又显出丰盛背后的精细。其中用到海鲜的就有一道巴戟杜仲海龙瘦肉汤，巴戟、杜仲均是中药，海龙又称杨枝鱼、管口鱼，是海马的近亲。制作时将巴戟、杜仲、海龙洗净，猪瘦肉洗净，切块，全部用料放入锅内，加清水适量，武火煮沸后，改用文火煲两小时，调味后饮汤吃肉。同样的主料，不同的药材，就能煲出口味、功效不同的汤，这种精细的搭配实际上也是

turnips, dried squids, eggs, lettuces, watercress, cabbages, minced meat, and abalones. Smooth and tender vermicelli roll with seafood, vegetables and meat brings a savory and appetizing taste which is beyond expression.

In the past, Lichee Bay in Guangzhou was a popular gathering place where many boats slowly steered to the shore carrying freshly cooked porridge to be sold to people returning home early in the evening. Over the years, Guangzhou has witnessed these scenes of the old days. Nowadays, the warm porridge on the boat has been given an official name —Tingzai Porridge (also called Mixed Seafood Porridge). Originally, the porridge was made of freshly caught fishes, shrimps, crabs, clams, and snails. In fact, boat owners put whatever they caught in the porridge, making it fresh, sweet, and savory. Today, Tingzai Porridge has its routine cooking method: first, cook a pot of porridge with japonica rice. Then add fresh shrimps (fillets), peanuts, jellyfishes, squids, bran, scallions, and gingers to mix and finally boil the porridge together with all the ingredients and seasonings just before it is served. A combination of seafood and freshwater fishes provides diners a multilayered long and protracted savory flavor.

In Guangdong where the climate is warm and humid, Cantonese are fond of making soups that can be considered as a millennium-aged mystique that preserves their health. Cantonese do cherish a genuine love for soups, especially double-stewed soup, which has almost become a necessity in the life of locals. There are hundreds of documented methods to make double-stewed soup and they all show that almost everything edible can

广府汤的秘密。

先上汤，后上菜，是粤菜宴席中的一个特色。粤菜号称选材广博，穷尽水陆之珍，"海八珍"自然不会遗漏。要说"海八珍"，就不得不提起著名的官府菜谭家菜。谭家菜源自同治年间进士谭莹对美食的偏爱。广东人谭莹酷爱珍馐美味，又好客酬友，常于家中作西园雅集，谭家菜自此发祥。后家道中落，谭莹不得已替人摆宴谋生，于是就有了"戏界无腔不学谭（谭鑫培），食界无口不夸谭（谭家菜）"的说法。

谭家菜"长于干货发制"，"精于高汤老火烹饪海八珍"，尤以海味菜最为著名。谭家菜对口味的追求也表现在选料精细，买熊掌需选味道鲜美的左前掌，买鱼翅需选有名的吕宋黄，买鲍鱼需选珍贵的紫鲍。例如鸡球鲍鱼一菜，将水发紫鲍四两用清水洗净，剞上方块花联刀，再斜切成厚一分半的片。将笋鸡肉四两切成一寸半见方的块，在每块鸡肉的四边略奇花刀，中间不要剞刀（目的是使鸡肉成球）。取两个鸡蛋去黄留清。用少许精盐、料酒、湿淀粉、白糖、味精、蛋清将鸡肉拌上味，腌渍几分钟。取一炒勺，注入花生油半斤，烧至六成热时，放入鸡肉炸之，待肉收缩成球形，捞出沥油。先将二两焖鲍鱼汤和三两老母鸡汤放入双耳锅内，煮

be added to the soup. In addition to the abundance of ingredients, the cooking process is significantly refined. Many questions need to be considered during the cooking process, for example, should the Chinese-style casserole be used uniquely in making the soup, at what temperature should the cooking be, how long should be the braising time after turning off the heat, at what time should different herbs be added, etc. Among all the dishes, pipefish lean pork soup with Radix Morindae Officinalis and Eucommia Ulmoides is made of seafood ingredient. Radix Morindae Officinalis and Eucommia Ulmoides are Chinese herbs , while "Pipefish, "also called cornetfishes, is a close relative of the sea horse. To make this soup, first, wash clean Radix Morindae Officinalis, Eucommia Ulmoides and Pipefish. Also wash clean pork lean and cube it. Put all ingredients into a pot and pour moderate clean water. Then put the pot over high heat to boil. Then reduce to low heat and simmer for two hours. The same ingredients, collocated with different herbs can create soups with different flavors and efficacies. Such fine collocation is, in fact, a mystique of Guangdong soup.

A main characteristic of Guangzhou banquet is that the soups are served first, and the other dishes are followed. Guangdong Cuisine is prominent for its wide and varied ingredients. It is a cuisine where diners expect to find every precious delicacy from both the land and the sea including the "Eight Seafood Delicacies. "Any discussion on "Eight Seafood Delicacies" cannot be complete without mentioning the Tan Family Cuisine, which is derived from Tan Ying (AD1846－1888), a native of Guangdong Province who secured a high position in the Qing court. Tan was known to have a passion for

开后加入鲍鱼及鸡球，滚煮十分钟，再加入酱油、淀粉、精盐、料酒，调成浓汁，淋上鸡油，出锅装盘即可。成品软烂鲜嫩，汤味浓郁鲜美。

在潮汕地区，也有一道颇显丰盛的传统名菜——什锦乌石参。主料为乌石参，辅料有鸡肉、肥膘肉、蟹肉，香菇、莲藕、冬笋、干贝、火腿，另有葱、姜等调料。制作时乌石参内面切粗花刀，然后将乌石参过滚水三分钟，捞出后入蒸笼蒸一小时。干贝加上汤入笼蒸二十分钟备用。将鸡丁、肥膘丁投入滚水，把鲜菇丁、鲜藕、笋花丁、火腿丁，蟹肉等一起泡熟后倒入笊篱。取出蒸笼内的海参，倒去原汤，再将干贝与处理好的什锦丁料倒入海参肚里，然后翻入大汤碗。最后用干净的锅加上汤，加调料调味，将上汤淋入翻扣过的海参即成。这道菜虽以乌石参为主料，但品尝时，乌石参包裹的多种新鲜食材的原味混合在一起，味蕾上的感受是奇妙的。什锦带来的则是食材与味觉上的双重丰盛。

除了食材的丰盛，广东人在吃法上也有独创，比如历史名菜甲子鱼丸。据载，南宋末年宋帝赵昺与陆秀夫军败，退至陆丰市甲子镇待渡时，被当地的义军捉住，惊恐万状的宋帝见对方不是敌人，而是大宋百姓，便道出了真实身份。当时的义军首领郑复翁，连忙让妻子巧姑做当地名肴"鱼

delicacies and he spent much of his time hosting lavish and elegant banquets. However, his career and his financial situation worsened later. So, he had no choice but to make a living by arranging banquets for others. His exquisite skills even gave rise to the folk saying: "In the world of Beijing opera, everyone learns from Tan (for Tan Xinpei, the opera virtuoso); in the world of gastronomy, everyone praises Tan (for Tan family Cuisine). "

Tan family Cuisine is known for its seafood dishes, especially the process of soaking dehydrated seafood (the process includes soak-boil-soak until the dehydrated food is edible) and cooking the "Eight Seafood Delicacies" with soup-stock on just-right heating temperature. The pursuit of taste of Tan family Cuisine is also reflected in their fine assortment of ingredients. Nothing other than rare and expensive food satisfies their demands. For instance, the left forepaw is the most savory and valuable part of a bear paw, the famous Luson shark's fin that is alleged to be imported from the Philippines, and the precious purple abalone that has pearly luster on the back.

To cook the dish braised abalone with chicken balls, first, take 200 grams of soaked abalone and wash clean, carve the abalone into the shape of cubes and cut them into slices of 1.5cm from 45-degree angle. Second, take 200 grams of chicken and shape them into balls, then crack two eggs, separate the yolk from the egg white. Mix moderate salt, cooking wine, wet starch, white sugar, MSG and white egg with chicken to marinate for a few minutes till it is tasty. Third, add 250 grams of peanut oil to the wok, add chicken and fry in hot oil over 150 degree heat. Then remove and drain it allowing

丸"，热情款待宋帝。饥肠辘辘的宋帝面对碗中上下翻滚的颗颗洁白鱼丸，品尝后觉得鲜美无比，龙心大悦，声称"甲子鱼丸"有救驾之功。自此"甲子鱼丸"被视为名菜之首。甲子鱼丸的制作看似简单，其实极其讲究，要求选料精、原料鲜，一般只选取鲜黄鱼、马鲛鱼、鳗鱼等。制鱼丸时，先取鱼青，鱼青忌有红肉和鱼骨，再剁成肉泥，加调料搅拌。后将盆中的鱼肉用手猛力搅打，加入盐水，再打至鱼胶放入冷水中能浮起为准。可见，鱼丸制作需要内心有着一份精细。

世界上许多国家的菜肴都有着丰盛的选材，并会使用大量海鲜。可若论起食材之丰盛，制作之精细，可以与粤菜比肩的外国菜系并不多见。但这并不妨碍外国美食对于丰盛的表达，比如地中海的名菜——马赛鱼汤。官方的《马赛鱼汤指南》中，需以岩鱼烹制汤底，再以海鲂、蝎子鱼、鲅鳒和海鳗作为主要食材熬制。在民间则会加入贻贝、螃蟹等，期间还需要加入多种香料，并有过滤工序。这些材料口感和营养成分各不相同，但汇聚之后却丝滑馥郁，味道鲜美。在濒临大西洋的葡萄牙，丰盛则表现为最正宗的海鲜锅。葡萄牙海鲜锅通常以蛤蜊为主料，再加入番茄、洋葱等蔬菜和一些香料，煮一刻钟后再加入海虾、章鱼、

the chicken to shrink to the shape of balls. Fourth, pour 100 grams of braised abalone soup and 150 grams of chicken soup to the casserole and boil. Then add abalone and chicken balls and boil for ten minutes. After that, add soy sauce, starch, salt and cooking wine to stir until the sauce thickens. Finally, sprinkle chicken oil to finish. The meat is tender and soft while the soup is thick and savory.

Assorted sea cucumber, as a traditional dish in Chaoshan region, also shows the abundance of ingredients. The main ingredient is sea cucumber while other ingredients include chicken, pork fat, crabs, mushrooms, lotus roots, winter bamboo shoots, dried scallops, and ham. The seasonings include scallions and gingers. To cook the dish, first, score on the inner face of the sea cucumber, then blanch the sea cucumber in boiling water for three minutes then steam it for one hour. After that, add clear soup into the dried scallops, steam for twenty minutes then set it aside. Put diced chicken and diced pork fat together with diced fresh mushrooms, fresh lotus roots, diced bamboo shoots, diced ham, and crab meat into the boiling water. Boil until they are well done, then remove and put them in a strainer. Take out the sea cucumber and pour the original soup away. After that, stuff dried scallops and above assorted ingredients inside the sea cucumber and move them to a big bowl full of soup. Finally, pour the soup onto a clean casserole, then add seasonings as well as the well-soaked sea cucumber to finish. Though the main ingredient of the dish is sea cucumber, various flavors of ingredients wrapped inside the sea cucumber blend with each other, providing diners with a feast in both smell and taste.

Guangdong people are not only proud of abundant ingredients, but also take pride in many self-created dishes. One of them is the famous historical dish, jiazi fish balls. According to records, in the late Southern Song Dynasty when the country was unstable, Emperor Zhao Bing (reigned from AD1272 to AD1279) and the prime minister Lu Xiufu (AD1236−1279) lost the battle and fled to Jiazi town in Lufeng city. They were, however, accidentally caught by the righteous army that was composed of common people. Seeing that they were not enemies, the emperor breathed a sigh of relief and revealed his true identity. Zhen Fuweng, the leader of the righteous army, told his wife to serve the famous local dish fish balls for the emperor who, by then, was as hungry as a wolf. The dish was several white fish balls rolling up and down in the bowl. They tasted savory and delicious enough to make the emperor content and happy even though he was on the run. Fish balls was then acclaimed as the dish that had rescued the emperor, pushing it atop the ranking of dishes. The cooking method of the dish looks simple but, actually, it is complicated because it involves fresh and exquisite ingredients like yellow croaker, mackerel, and sea eel. First, get minced fillet without fish bones, add seasonings and stir well. Then, beat the rest of the fish in the basin and add saline water. Continue to beat the fish until it turns into fish gelatin that can float on cold water. Significant care needs to be taken when cooking this dish.

Although the dishes of many countries in the world have a wide selection of ingredients and use a load of seafood, they pale in comparison with Guangdong Cuisine's abundance in ingredients and refined cooking methods. However, they deliver "abundance" in different ways. For example, Bouillabaisse is a famous dish

扇贝、螃蟹、生蚝等，最后慢慢熬制而成。一锅食材的汇聚，带来的是最浓郁的海洋味道。这大概就是不同地域的人们对于丰盛的理解吧。

今天，越来越多的广东人走向全球，带动了粤菜的传扬，使得粤菜在世界餐饮中占据了日益重要的位置。在某种程度上，粤菜已经成为中国美食在世界上的名片。也因此，粤菜丰盛背后的精细，精细背后的文化，正被世界上越来越多的人接受和欣赏。相信在世界各国交融发展的过程中，粤菜的丰盛与精细会愈发具有生命力。

originating from the Mediterranean. According to the official Guidebook of Bouillabaisse, this dish is cooked by first preparing a pot of soup based on Rascasse. Then add John Dory, Red mullet, Monkfish and Congereel as main ingredients which are then cooked together. Common people are apt to use whatever seafood is available, especially mussels and crabs served with various spices. Although these ingredients have different tastes and various nutritive value, there is a harmonious blend of flavors which provides a silky and tender taste. In Portugal, which is close to the Atlantic Ocean, we can find "abundance" in an authentic seafood pot. Portuguese seafood uses clams as its main ingredient and supplemented by vegetables like tomatoes and onions as well as some spices. After fifteen minutes of boiling, shrimps, octopuses, scallops, crabs, oysters are added and stewed slowly until they are done. The convergence and integration of all the ingredients make up the charm of the sea. As "There are a thousand Hamlets in a thousand people's eyes, "the comprehension of "abundance" varies from people of one region to another.

Guangdong Cuisine' prominence outside China is partly due to the large number of Cantonese that migrate abroad. Besides occupying an increasingly important position in the world of food and beverage, Guangdong Cuisine has, to a certain extent, become a calling card for Chinese Cuisine all over the world. Therefore, more and more foreigners have begun to appreciate the "refinement" and "abundance" of the cuisine, and more importantly, the culture behind it. It is believed that Guangdong Cuisine and its vitality in abundance, will be elevated unto the world platform.

自然，琼崖的风情

第八章

Hainan Cuisine:
Natural Flavor

　　海南省，是中国最南端的省份，古称"南服荒缴"之地。因远离中原，自古以来朝廷难以顾及，直至西汉武帝时始归中央政权管辖。光绪时，升崖州为直隶州，与琼州并列，故此后海南又称琼崖。一九八八年，中央政府撤销广东省海南行政区，设立了今天的海南省和海南经济特区。

　　海南地处热带，气候炎热，加之四面环海，海鲜、野味、禽畜和果品等十分丰富，故海南人饮食立足于当地特产，讲究鲜活为主，味重清鲜，原汁原味，且注重养生。海南菜已有两千多年的历史，发源于中原餐饮，融汇闽粤烹饪技艺，吸收黎苗山野气息，又兼有东南亚风味，是今天中华饮食中富有活力而又特色鲜明的地方菜系。综合来看，海南菜主要具有三大特色：一是取料独特，二是水果入菜，三是风味小吃别具一格。

　　因地处热带，海南省的物产别致，是其他省份所不及

Hainan Province, the southernmost province of China, is known historically as "region in the bleak south borders." It is so far away from the Central China that since ancient times, very few dynasties had jurisdiction over the region until 110BC when Emperor Wu of the Western Han Dynasty formally established a military garrison there. Under the reign of Emperor Guangxu in the Qing Dynasty, Yazhou Prefecture was upgraded to Yazhou Municipality, which, together with Qiongzhou Municipality, are collectively known as "Qiongya," a nickname of Hainan Province. In the memorable year of 1988, the central government abolished Hainan administrative region of Guangdong Province, paving the way to the establishment of the current Hainan Province and Hainan Special Economic Zone.

Located in the tropics and surrounded by sea, Hainan Province brings its residents into closer contact with nature in which seafood, game, livestock, and fruits are plenty and easily acquired. Taking full use of local fresh ingredients, Hainan Cuisine is light and savory and it especially emphasizes on preserving its original flavor and never forgetting its pursuit of health. It is thanks to over two thousand years of development that Hainan Cuisine integrates the food culture inherited from the Central Plains, the exquisite cooking skills from Fujian and Guangdong cuisines, the natural flavor of the Li and Miao nationalities, and the taste of southeast Asian food. All of these make Hainan Cuisine a vivifying and distinctive local Chinese Cuisine with three distinctive characteristics: unique selection of ingredients, use of fruit in

cooking, and Hainan-style flavorful snacks.

Hainan Province is remarkable for its rich and distinct ingredients due to its unique climate. It boasts Hele crabs from Wanning county, sea snakes from Yazhou district, Pineapple sea cucumbers from Xisha and Nansha islands, coconut crabs, frogs crab and an endless list of other products. The dishes with these ingredients are too many to count, so we may only mention a few of them at this juncture.

In Wanning county, Hainan Province, there is a town named Hele. It is said that in the Tang Dynasty, people migrated from Putian city in Fujian Province to Hele village which lies to the west of Gangbei continental sea. The newcomers made special crab nets to catch crabs as a way of making a living. The crabs they caught from the gulf of Gangbei were such flavorful delicacies that locals not only named the crab "Hele crab," but also chose the crab dish as one of the four most famous dishes in Hainan Province. Soon the crabs were sent to emperors as a tribute and in this way, they gained popularity around the entire country. Centuries later, under the reign of Emperor Guangxu in the Qing Dynasty when all the governors in Hainan paid tribute to high officials by sending them Hainan specialties, Tangli, an official under the magistrate of Wanzhou Prefecture, in spite of the difficult journey, had made all the way from Hainan to Beijing and sent "Hele crabs" to imperial princes and court ministers. Hardly had they tried the crabs when they started lavishing praises on this dish: "The Hele crabs are quite aquatic treasures."

There are many ways of cooking Hele crabs and one of

的，这为海南菜提供了丰富而又独特的食材，比如万宁县的和乐蟹，崖州的海蛇，西沙、南沙群岛的梅花参，以及椰子蟹、琵琶蟹等。以这些食材为主料的菜品也颇多，这里就略说一二。

和乐本是海南省万宁县的一个镇。据传，唐时，从福建莆田迁居港北小海西面的和乐村渔民，专门制作一种蟹网，在港北的海湾之中捕捞肥蟹，因风味独特，遂以"和乐蟹"称之。作为海南四大名菜之一的和乐蟹，早在唐朝就已十分出名，还曾被列为"贡品"进贡朝廷。相传，清朝时期就有"唐立送蟹"的故事。清康熙年间，海南各地官吏均进贡海南特产给朝廷高官，万州（万宁）知县派员（唐立）将万州海鲜特产——和乐肥蟹千里迢迢"进贡"，品尝了和乐蟹的朝廷官员，纷纷赞不绝口，称"和乐肥蟹乃水产之珍品也"。

和乐蟹烹饪方法多种多样，最常见的是清蒸，在烹饪时选取雌性和乐蟹数只，洗净入锅，加适量清水、精盐、姜片、葱片，旺火蒸十五分钟后取出。洗净黏附物，在明火上稍烤干蟹身，横刀切为两半，再下十字刀切块，摆盘整齐即可。上桌时配以蒜蓉、姜末、糖、盐、鲜橘汁、辣椒丝调成的佐料蘸食。此菜以清蒸之法保持食材的自然风味，其成色金黄，香味扑鼻，且营养丰富。

the simplest ways is to steam them. First, choose several female crabs and wash clean. Then steam them for fifteen minutes over high heat after adding water, salt, sliced ginger, and scallions. Next, clean the crabs again and grill them over high heat for a while until they dry. Next, cut them in half horizontally from the middle and then cut crosswise into chops. Finally, arrange them side by side on a plate. Mix well minced garlic, bruised ginger, sugar, salt, fresh juice and shredded chili to serve as the dipping sauce. By steaming the crabs, significant care is taken to preserve the original flavor of the ingredients. This dish is golden in color, it is savory and rich in nutrition, and it has a long-lasting and encompassing smell.

Sanya city, historically known as "Yazhou," is the birthplace of sea snakes. Locals, therefore, created a famous dish with sea snake and chicken and named it Dragon Meets Phoenix (indicates extremely good fortune). To cook the dish, first, wash clean the snake, blanch it in boiling water, and remove its bones. Next, remove the chicken bones and cube them. Then coat them with egg white and starchy flour. After that, put garlic cloves in a wok where oil is being heated and fry the mix until it turns brownish. Fry sea snake and cubed chicken and then remove them to drain. Again, put the gingers and scallions into the wok containing heated oil then add sea snake, chicken, as well as a little cooking wine and soup. Then combine MSG, oyster sauce, dark soy sauce, and salt to braise for five minutes. Thereafter, cube the snake and arrange them on the bottom of the bowl ensuring to keep the original shape of the snake (with the snake skin down). Cover it with cubed chicken

三亚市古称"崖州"，盛产海蛇。当地人以海蛇与鸡肉创制了一道名菜，谓之曰"龙凤和合"。制作此菜时先将整条海蛇洗净，带皮放至热水中余烫，取出骨刺，然后将鸡肉去骨切块，上蛋白，醮生粉。猛火烧锅下油，入蒜瓣炸至焦黄，下海蛇、鸡块过油后，捞出沥油。再次热锅放少许油，入姜葱爆香，复入海蛇、鸡块，点黄酒，加上汤，调入味精、蚝油、老抽、精盐，文火焖五分钟后取出。后将海蛇切块，按原状摆置碗底（蛇皮朝下），再铺上鸡块、油炸蒜瓣，整碗反扣入瓦煲中，加入原汁，盖上，置炭中火煨至酥软，调入芝麻油、胡椒粉，原煲上席即可。此菜蛇肉入口肥厚细嫩，味道鲜美，风味独特。

　　海参、鲍鱼、鱼翅，俗称"崖州三珍"。三亚和西沙、中沙、南沙群岛海域盛产的海参种类颇多，主要有梅花参、花刺参、黑乳参、红腹参等。国宴上用的梅花参就是西沙、南沙群岛海域所产。海参的食用方法很多，可用鸡汤文火炖，也可切片加火腿、笋尖等佐料清炒，还可以制成甜品：用海参和鸡蛋，加上桂圆、冰糖等一起清炖，也是一种具有滋补功效、营养丰富的佳肴。

　　热带季风气候带来的光照和雨水使得海南盛产椰子、芒果、菠萝、杨桃、木瓜、荔枝等上百种热带瓜果，堪称水果

and brown garlic cloves. Next, place the whole bowl upside down and move the food into a clay pot, then pour the original soup into the pot. Next, cover and cook over medium heat until the meat is tender. Finally, top with sesame oil and pepper and serve it in the clay pot. Given that the snake meat is plump and tender, the dish is savory and flavorful.

Sea cucumber, shark's fin, and abalone are known as "Three Treasures of Yazhou." There is a rich varieties of sea cucumbers produced in Sanya, Xisha, Zhongsha, and Nansha islands and they mainly include Pineapple sea cucumber, Golden sea cucumber, Black sea cucumber, and Edible sea cucumber, of which Edible sea cucumber, produced in Xisha and Nansha islands, has made its way onto the State menu. Chefs need especial cooking skills to prepare this dish. Whether it is cooked by steaming with chicken soup over low heat or fried together with ham and tender tips of bamboo shoots, the sea cucumber is consistently savory and delicious. When sea cucumbers and eggs are steamed together with longan and rock candy, they constitute a special dessert full of nutriment and good for the health.

Endowed with sufficient sunshine and rain, Hainan Province is considered as the paradise of fruit. As a matter of fact, there are hundreds of tropical fruits such as coconut, mango, pineapple, carambola, papaya, and litchi on this island. In order to take full use of these gifts from nature, residents have created a considerable number of fruit-based dishes including Coconut Chicken, Papaya Pot, Sautéed Shrimps with Mango Juice, Yanglan Sour Fish Soup, and Hainan Coconut Boat. The ingredients used in cooking Sautéed

的天堂。所谓"物尽其用"，海南菜便有了"水果入菜"这一特色，如椰子鸡、木瓜盅、芒果汁淋虾、羊栏酸鱼汤、海南椰子船等等。芒果汁淋虾以芒果和海虾为主料，先将虾仁洗净，加少量盐、胡椒粉、黄酒腌制，青瓜切块，红椒切片。其中芒果做两用，其一取果肉切丁，其二取果肉加水制芒果汁。锅中加油烧热，放入虾仁煸炒，入青瓜和红椒，加盐调味，再加入芒果丁翻炒，最后淋上芒果汁即可。芒果丁做辅料保留口感，芒果汁做浇汁提味，芒果与海虾相融合，既保持了芒果的清香，又有了虾的鲜香，实在是一道色香味俱全的佳品。

羊栏酸鱼汤是三亚回族的传统美味，因当地回族人聚居羊栏镇，故得此名。这道菜的食材取自三亚近海各类杂鱼，辅以酸豆、酸杨桃、西红柿等，煮出来的鱼汤新鲜且没有腥味，不但受回族人民喜爱，其他各族人民特别是到三亚旅行的游客也十分青睐。其制作方法十分朴素，只需将水煮开，把洗净的近海鲜鱼放入锅中，然后依次放入酸豆、酸杨桃和西红柿等辅料调味，加少许油、黄酒、白胡椒粉、醋和生抽，盖上锅盖煮至鱼熟汤美时，放入盐调味，即可食用。虽然制作方法简单，但成菜鱼肉鲜嫩可口，汤汁酸咸爽口，令人胃口大开。

Shrimps with Mango Juice are mango and shrimps. To cook the dish, first, wash clean shrimps, then add a little salt, pepper, and cooking wine to marinate the shrimps. Cube green cucumbers and slice cayenne peppers and next, dice half of a mango and make mango juice with the other half. After that, put the shrimps into a heated wok with oil and sauté. Then add green cucumbers and cayenne peppers as well as a little salt to mix well. Also, add diced mango to sauté quickly and finish by drizzling mango juice. The punchline of the whole process is the magical use of mango, with the diced mango serving as the ingredient that retains the freshness while mango juice is the topping. The dish is a perfect blend of mango and shrimp flavors. It has a beautiful color and a special taste.

Yanglan Sour Fish Soup is a dish named after Yanglan county, a place inhabited by the Hui people. As a traditional delicacy in Sanya city, the dish is popular among not only the Hui people, but also people of other ethnic groups, especially visitors to Sanya. Although this dish is cooked mainly using special kinds of different fishes that abound in the coastal area, other ingredients including capers, sour carambola, and tomatoes are indispensable to make the soup delicious. To cook the dish, first, clean the fresh fish and then put it into the wok with capers, sour carambola, and tomatoes. After that, add a little oil, cooking wine, white pepper, vinegar, and light soy sauce then cook until it is done. To finish, add some salt. Despite its simple cooking method, the dish is mouthwatering given its tender fish and sour but savory soup.

The charm of Hainan Cuisine lies not only in its unique

除了取料独特、水果入菜以外，海南的风味小吃也别具一格，诸如椰叶薏粑、珍珠椰子船、煎粽、清补凉等等，种类繁多，花样百出。其中以海味入菜的更不在少数。海南儋县新英镇、白马井镇等地渔民每年出海捕捞大量的红鱼，因无法及时食用完毕，又苦于气候炎热，于是便把红鱼加工成鱼干，其中新英镇鱼干尤其闻名。渔民将红鱼从鱼背至鱼头直线切开，底腹和下巴却不切断，掏出内脏后置于盐水池中浸泡，几天后取出晒干即可。后来，海南疍民又将鲜美的红鱼干与醇厚的五花肉相结合，创制了疍家咸鱼煲。此菜烹制简单，只需先将红鱼干泡发切片，然后将码味后的五花肉以文火烧十分钟后，放入红鱼干，以瓦煲焖熟即可。其成菜鲜香浓郁，美味无比，是疍民佐酒下饭的美味佳肴。

同样以红鱼为主料的还有儋州风味小吃——红鱼粽。不过这种红鱼粽并非米做的"粽子"，而是一种酶香鱼。红鱼粽不仅在海南受欢迎，在东南亚也深得消费者的青睐。比起红鱼干，红鱼粽的制法相对复杂：先将四斤左右的红鱼去鳞、除肋和内脏，把食盐塞入鱼腹，置于腌鱼池或鱼桶中，鱼身再抹上食盐，腌制十日左右取出晒干；而晒制时也有很多讲究——鱼粽不能吊晒，必须每条分开排列晒，每晒两个小时需翻身一次，中午太阳强烈时要晾至阴凉处，使其通风散热。

selection of ingredients and its use of fruit, but also in its variety of snacks including glutinous rice dumplings covered with coconut leaves, Pearl Coconut Boat, Hainan Fried Rice Dumplings, and Refreshing and Nutrient Herb Soup. The list of Hainan snacks is endless. Seafood-based snacks are undoubtedly not in the minority in this area. Fishermen of Xinying county and Bai Majing county in Zhanzhou city catch a large number of red fishes every year. These fishes, however, can neither be eaten up in time, nor can they be stored easily due to the hot weather. Therefore, they have been processed into dried fishes including the dried fishes of Xinying town that are particularly famous. Fishermen cut the red fish along a neat straight line from back to head making sure not to completely separate the head and the chin in the process. After removing the viscera, soak the fish in a salt basin for several days and then leave it under the sun to dry. Later, the Dan people (people who spent all their life on the boat) in Hainan Province invented the dish Dan family salted fish hot pot, which was a wonderful marriage between the delicious dried fish and the plump streaky pork. To cook this dish, first, soak the dried fish then slice it. After that, braise the marinated streaky pork for ten minutes over low heat and add dried fish to braise together until they are well-down. The dish is savory and tasty and it is the main dish of Dan people.

Red fish also serves as the main ingredient of another local snack in Zhanzhou city—red fish dumplings, which attract people not only in Hainan Province but also in the Southeast Asia. It is worth mentioning that the "dumplings" referred to here are, in

fact, the enzymatic salted fishes rather than the one that is made of rice. The steps involved in making this dish are more complicated than those involved in cooking dried red fishes. First, scale the red fish (about two kilograms) and remove the ribs and viscera. Then put salt into the cavity and over the fish. Place the fish in a basin to marinate for about ten days and then dry it under the sun. It is during the sun-drying process that people need to pay attention to several tips. First, the fish must be separated from one another and perfectly arranged in order. In particular, the fish should not be hung to dry. Second, turn the fish over after every two hours. Third, take the fish to a shaded place to ventilate and dissipate the heat at noon when the sunlight is too strong. In addition, before the drying process, rough straw paper should be stuffed into the mouth of the fish and the fish head should be wrapped in white paper to prevent flies from eating it. Once all these steps are completed, the fleeting time helps to make red fish dumplings. The only way to cook the dish is through

另外，在晒鱼前要将草纸塞进鱼口中，用白纸将鱼头包好，防止蝇虫进入。此后，只需时间赋予红鱼神奇的变化，风味奇特的红鱼粽就制成了。红鱼粽只能蒸食。取食之际就可闻到红鱼粽的酶香，将红鱼粽洗净切成小块，放入盘内，配以猪肉、姜丝等，下锅蒸制即可。蒸熟以后，揭开锅盖，更是香味四溢，其口感爽脆，令人难忘。

同样以"粽"为名的海南煎粽则是真真切切的米粽了。煎粽是流行于海口的地方风味小吃，与寻常粽子不同，无需粽叶包裹，只需煎熟食用即可，且其制作与食用不受节令限制，一年四季皆可。海南煎粽以糯米、莲子、冬菇、干虾米、干贝、叉烧等多种食材为原料，制作时先将糯米洗净，蒸熟待用。将莲子煮熟，脱皮除心，冬菇浸透洗净，滚水烫过后切成粒状，干虾米、干贝洗净蒸过，叉烧剁成粒状。将上述原料与熟猪油、味精、白糖、精盐等一起放入糯米中拌匀，捏成大小相同的饭团。热锅下油，将糯米团均匀涂上鸭蛋清，放进热油中小火煎炸，不断翻转使之均匀受热，煎成金黄色即可。做好的煎粽色泽金黄，外焦里嫩，米香中裹着莲子的清香和海味的鲜美，十分可口。

气候类型为热带季风的地区基本都位于亚洲，除中国外，大多分布于中南半岛、印度半岛与菲律宾群岛北部，其海鲜

steaming. First, wash clean and cube the fish, then put it in a platter and add pork and shredded ginger to steam until it is completely done. As soon as you lift the lid, you would be greeted by a savory smell and the crisp taste will linger for a long time in your mouth.

Hainan fried rice dumplings, another popular local snack in Haikou area, resembles the red fish dumplings in name, but differs in ingredients especially because the "dumplings" are just made from rice. However, the fried rice dumplings are not the same as the ordinary rice dumplings. They are not wrapped with reed leaves and they are available all year round. The main ingredients are glutinous rice, lotus seeds, winter mushrooms, dried shrimps, dried scallops, and roast pork. To cook the dish, first, wash clean the glutinous rice and steam until it's well cooked. Then boil lotus seeds and remove their skin and plumule. After that, soak the winter mushrooms and wash them clean. After blanching in boiling water, dice them. Then wash clean dried shrimps and dried scallops and steam well. After that, dice roast pork and put them in a bowl, combine glutinous rice with the above ingredients and add lard, MSG, sugar, and salt. Stir well then pinch them into equal-sized rice rolls. Thereafter, coat the rice rolls with duck egg-white and fry them in a wok over low heat. Continue turning and frying until they turn golden brown. The dish is crisp outside and tender inside and is a perfect and harmonious blend of the flavor of rice, lotus seeds, and seafood.

The tropical monsoon climate affects Asian countries including China, the Indo-China Peninsula, the Indian Peninsula, and the northern part of the Philippine Islands. People who live in

菜肴以西式烹饪法为主。在泰国曼谷，可以品尝到红咖喱椰汁煮虾，在印度孟买，则有玛萨拉煎鱼、冰罗勒蒜香鱿鱼等，其中玛萨拉酱与香料罗勒是当地的特色。在孟加拉的达卡，在较为高档的海鲜饭店中，白汁沙司贻贝和香煎红鲷鱼较受欢迎。在菲律宾的马尼拉，椰汁醋鱼沙拉则是有名的菜肴。这些地区海鲜菜肴种类不多，但菜肴中香料倒是一大特色。

　　海南是个海洋群岛省份，岛屿和海洋是自然的馈赠，而风情则是生活在这片土地上的人们所塑造的。涛声阵阵，椰林荡漾，海南菜这一仍然年轻的菜系，将带着闽粤的技艺、民族的气息、异域的风味，继续捡拾着它自然丰硕的果实，释放出它原汁原味的独特风情。

these regions opt to cook seafood in the western way. For example, you can try the delicious dish boiled shrimps with red curry and coconut juice in Bangkok, Thailand. In Bombay, India, it is not difficult to have a taste of their local food including Marsala Fried Fish, Squid with Garlic-flavored Ice Basil. Marsala sauce and basil are the main ingredients. of these delicious dishes. While in some fancy seafood restaurants in Dhaka, Bangladesh, the dish Mussels in White Sauce and Fried Red Snapper are more popular. In Manila, Philippines, fish salad with coconut juice and vinegar is another example of a tasty flavor. The number of seafood dishes in these regions may be limited, however, their richness in terms of the variety of spices cannot be overemphasized.

Hainan is a province composed of many islands. The natural environment and the fertile ground contribute to the rich food resources that have enabled people to create their own romantic flavor known as Hainan Cuisine. Through all these seasons, Hainan Cuisine has developed upon the exquisite skills of Fujian and Guangdong cuisines that combine an ethnic culture and exotic flavor. In the future, this young cuisine that is growing with the rolling waves and dancing coconut trees will continue to harvest nature's gifts and to show its romantic flavor to the world.

荟萃，港澳的融合

第九章

Hong Kong and Macao Cuisines:
Integrated Constellation

　　对于一座充满活力的城市来说，前进的脚步往往是匆忙的。人们追逐梦想，迈向未来，<u>丝毫不敢懈怠</u>，而在匆忙之余，美食就成为最好的慰藉。对于生活在香港和澳门的人而言，美食就是这样的存在，是一种与发展结合在一起的精彩。

　　香港和澳门自古以来就是中国的领土。两个特别行政区分别位于珠江口东西两侧，隔江相望。香港是一座高度繁荣的国际大都市，有"购物天堂""美食天堂"之称，而澳门是世界旅游休闲中心，也是世界四大赌城之一。究其渊源，历史上两座城市都曾归于广东治下，因此其居民多为广东人，也有不少邻省移民。但自十六世纪中期澳门被葡萄牙人占据以后，西方的饮食方式尤其是葡人的饮食方式便随之传入澳门。而香港在十九世纪被英国占领以后，其传统的饮食结构也发生了翻天覆地的变化。二十世纪末，香港和澳门先后回归中国并被划为特别行政区，由此，两地进入新的时代。对

P eople living in vibrant cities are generally busy pursuing their dreams and never dare to slow down. However, at the moment, when they are overwhelmed by various chores, delicacies seem to become the magic that dissipates their fatigue and relieve their tension. This is even truer for dwellers of Hong Kong and Macao who consider delicacies both as a vitamin for energy boost and a comfort that provides them with peace of mind amidst the rapid development of society.

Hong Kong and Macao have been territories of China since ancient times. These two special administrative regions are located respectively at the east and west sides of the Pearl River Estuary, facing each other. Hong Kong is a popular international metropolis known as "shopping paradise" and "gourmet paradise, "while Macao is a world tourism and leisure center and it is home to one of the world's four largest casinos. Such prosperous and multicultural cities have their own stories of joy and sorrow. Historically, the two cities were once under the jurisdiction of Guangdong Province. Their residents, therefore, are mostly Cantonese as well as some migrants from neighboring provinces. They lived peacefully until the middle of the 16th century when Portuguese troops invaded and occupied Macao. After that, western diets were introduced in the city. About three centuries later, Hong Kong became a colony of the British Empire after the Qing government ceded Hong Kong at the end of the First Opium War in AD1842. The cession had a great impact on the traditional diet in Hong Kong. At the end of the 20th century, Hong Kong and Macao returned to China successively and were designated as special administrative regions. Since then,

于香港和澳门而言，新时代是开放的，也是历史的，有本土文化，也有着世界文化，荟萃与融合成为代表性的词语，海鲜美食同样如此。

港澳濒临南海，拥有天然良港，渔业十分发达，故两地居民皆嗜食海鲜。香港海鲜酒家众多，食用海鲜是当地生活中不可或缺的部分。说到香港的本土菜，就不得不提起盆菜。盆菜是广东和香港新界的饮食习俗，是一种杂烩菜，已有数百年历史。盆菜本是源自新界围村习俗"打盆"，当地村民只在乡族的重要事件如嫁娶、添丁、满月等日子才烹制与享用盆菜。如今，盆菜早已不再限制于红白喜事，人们在逢年过节时也吃盆菜，盆菜也具有了喜庆团聚的意义。

关于盆菜的起源还有一个故事。据说南宋末年，宋帝为摆脱金兵追赶，落难至今天的香港元朗，正当随从四处张罗食物之际，当地村民得悉皇帝驾临，为表心意，纷纷将家中储备的猪肉、蔬菜以及新鲜捕捞的鱼虾等贡献出来，仓促间以木盆充当器皿，献到皇帝面前。

盆菜看似粗糙，实则考究，从选料上来讲，并没有特别规定，但一般都会包括鱼丸、土鱿、海虾、蚝豉、猪皮、萝卜、枝竹、冬菇、炸门鳝、鸡、鲮鱼球和炆猪肉等；从烹饪方法上来讲，各类食材分别要经过煎、炸、烧、煮、焖、卤

they have embarked on a new era. To dwellers of Hong Kong and Macao, "constellation" and "integration" mean that different cultures coexist and intermingle, and so are the seafood delicacies.

Situated near the South China Sea, Hong Kong and Macao boast natural harbors that contribute to their thriving fishing industries. Residents of both regions are so fond of seafood that it has become an integral part of their life. Seafood restaurants in Hong Kong are uncountable. As for traditional Hong Kong dishes, locals may think first of Poon Choi (basin dish), a chowder dish which has been a staple for people in Hong Kong and Guangdong Province for several centuries. Poon Choi is associated with the early settlers of the walled village, New Territories, who had the habit of serving the dish in large wooden, porcelain or metal basins. In the past, the dish was only served as an expression of village culture during celebrations connected with weddings, birth of baby boys, babies' first month of life, and other local events. However, the dish is now served at festivals to symbolize unification and happiness.

People in Hong Kong may be familiar with a folk tale about Poon Choi. According to the tale, during the late Song Dynasty, when Jurchen Jin troops invaded Song China, the young Emperor fled to the area around present day Yuen Long District of Hong Kong. To serve the Emperor as well as his army, the locals gathered all their best food including pork, vegetables, and freshly caught fish and shrimps. As they were doing this in a hurry, the villagers put the dishes in large wooden basins to present to the emperor.

Even though it does not look beautiful, the dish is complicated

等各种不同的制法后，方能入盆摆放；而从摆盆和食用方式来讲，盆菜的食物需按一定的次序一层层排好，上层要放些较名贵和需要先吃的东西，下层则放些容易吸收汁液的材料，而吃盆菜的时候也会由上至下逐层吃下去。如今，香港人食盆菜已成为一种时尚潮流，为迎合消费者的口味与要求，各大餐厅也纷纷推陈出新，海鲜盆菜、咖喱盆菜、素食盆菜、欧陆盆菜及日本盆菜等，百花齐放，使得盆菜更添一份魅力。

澳门曾经是个小渔村，本名为濠镜或濠镜澳，因当时泊口可称为"澳"，所以称"澳门"。又因澳门及其附近盛产蚝，因此又称作"蚝镜"。澳门当地有三种鲜蚝的吃法十分流行：其一，奶油蒜香烤蚝。将奶油入锅以小火熔化，加入蒜蓉、香菜末拌至溢出香味，熄火，撒上少许面包糠拌匀即成烤料。每只蚝上放半小匙烤料，入烤炉以中温烤五分钟即可，此法讲究保持蚝的鲜香。其二，鱼香蒸蚝。姜、蒜磨茸，葱切葱花，加辣酱、料酒、酱油、麻油、香醋拌匀。每只蚝上放半小匙调味料，入锅以大火蒸三分钟即可，此法以鲜嫩入味为上品。其三，生吃鲜蚝。加姜茸、葱花、海鲜辣酱，直接生食，讲究的是原生。

澳门地处咸淡水交界处，出产的蟹也种类繁多，且肉质鲜甜，因此以蟹为原料的菜品也颇多，比如有名的水蟹粥。

and exquisite to cook. Without specific rules, the dish ingredients though generally, includes ingredients such as fish balls, squids, sea shrimps, dried oysters, pigskins, radishes, bamboo shoots, winter mushrooms, fried eels, chicken, dace fish balls, and steamed pork. During the cooking process, different kinds of ingredients are used through various cooking methods such as fry, steam, boil, etc. In addition, particular attention needs to be paid to layering, which has a huge impact on the taste of the dish. By placing rare food on top and other materials capable of absorbing juice at the bottom of the container, it is easy for diners to try different ingredients that make up the dish layer by layer from the top to the bottom. As an international hub, Hong Kong can provide diners with a variety of flavors thereby helping Poon Choi to become even more attractive. There are now Seafood Poon Choi, Curry Poon Choi, Vegetarian Poon Choi, European-style Poon Choi and Japanese-style Poon Choi to fulfill different taste preferences of customers, whether they are locals or tourists.

Macao used to be a small fishing village named "Haojing" or "Hao jingao." Since at that time, the port was known as "Ao, "people therefore changed its name to "Ao Men" and also gave it the nickname "Oysters Jing" because the region, and its vicinity, were rich in oysters. Fresh oysters are definitely the favorite of the locals and there are three very popular oyster recipes: grilled oyster with cream and garlic, steamed oyster, and raw fresh oyster. Among them, grilled oyster with cream and garlic seeks to preserve the savory flavor of the oysters. To cook the dish, first, cook the cream over low heat until it

melts. Then add bruised garlic, bruised cilantro and stir until it gives off a savory smell. After turning off the heat, mix the above sauce with breadcrumbs and stir well to make sauce for grilling. Finally, mix every one-third tablespoon sauce with one oyster and grill over medium heat for five minutes. The second recipe, steamed oyster, seeks to preserve the tastiness and the tenderness of the oysters. After mincing gingers, garlics, and chopped scallions, combine them with spicy sauce, cooking wine, sesame oil, and vinegar then stir well to make it into the sauce. Finally, add half spoon sauce over each oyster and steam over high heat for three minutes. The third recipe focuses on the flavor of the oyster, so serve them raw with bruised gingers, chopped scallions, and spicy seafood sauce.

Macao, which lies at the junction of seas and rivers, is a harbor city where there are a great variety of crabs noted for their sweet and delicate flavor. Crabs are, therefore, widely used in Macao Cuisine as main ingredient including the famous dish— Macao Crab Congee, which, in fact, makes use of the most delicious part of three kinds of crabs—water crabs, male green crabs, and female green crabs—mixed with specially-made congee. This dish must be made by using the freshest crabs which are chopped into large chunks to prevent small pieces of shells from mixing in the congee. The congee is savory and delicious and has sweet crab flavor. In addition to Macao Crab Congee, there are other local specialties including Shrimp Bisque, Shark's Fin Noodles, Hengyou Fish Balls, etc.

Historically, Hong Kong and Macao were part of Guangdong

水蟹粥并非全以水蟹制成，而是取水蟹、膏蟹、肉蟹三种蟹的精华，再配上特制的粥所熬制。制作时需选最新鲜的蟹，还必须切大块，以防蟹碎壳混入粥内影响口感。其成品蟹黄丰腴，蟹肉鲜美，粥水蟹香浓郁。此外，澳门还有许多的本地特色美食，诸如鲜虾浓汤、鱼翅面、恒友鱼蛋等。

历史上港澳两地本隶属于广东，因此其饮食文化多受粤菜影响，有不少菜肴既保持了粤菜的特色，又融进了新的元素。名厨许沛荣便是在粤菜基础上创制出新菜式的代表人之一。许沛荣原是广东佛山人，两岁时同家人一起迁居香港，二十世纪六十年代开始在香港从事餐饮。八十年代，他在香港创办了"翠亨村茶寮"，与同事一起，创制出一大批全新菜式，人称"新派粤菜""港式粤菜"。这些新式粤菜还传到东南亚、美洲等地的中餐厅，在世界范围内产生了影响。他创制的港式粤菜包括串烧牛柳、上汤焗龙虾、雀巢牛柳丝等等。

其中，上汤焗龙虾是一道色香味俱全的海鲜名菜。制作时先将龙虾肉洗净剁块，粘上生粉备用。热锅下油，放入龙虾炸至七成熟后捞出。另起锅下油，入姜葱煸香，烹料酒，加上汤、精盐、白糖、胡椒粉调味。最后放入龙虾，勾芡装碟即成。此菜成品色泽鲜明，味道清淡鲜美。龙虾吃完后拌

Province, so their cuisines have been significantly influenced by Guangdong Cuisine. However, the cuisines in these two cities have also absorbed the quintessence of other cuisines around the world. Many chefs from this region, including the famous chef Xu Peirong, are proficient in creating new dishes based on original Guangdong Cuisine. Foshan city in Guangdong Province is the birthplace of Master Xu who migrated with his family to Hong Kong at the age of two. In the 1960s, he started his catering business, and in the 1980s, he founded Tsui Hang Village Restaurant Limited. In his restaurant, he, together with his colleagues, created a multitude of new-style dishes known as "New-style Guangdong Cuisine" or "Hong Kong-style Guangdong Cuisine." These new-style dishes are now spread all over the world. Today, one can easily find dishes like Skewered Sliced Beef, Lobster Poached in Chicken Consommé, Nestle Shredded Beef, etc., in Chinese restaurants in Southeast Asia and America.

The famous seafood dish Lobster Poached in Consommé is both appetizing and nice-looking. To cook the dish, first, wash clean lobsters and chop them up into chunks. Then after coating them with cornstarch, put them in a heated wok and fry them until they are medium cooked. Then, fry scallions and ginger and add cooking wine, clear soup, salt, sugar, and pepper. Finally, put lobster chunks into the wok and thicken the sauce with starch. The cooked dish is bright in color and it is savory and mellow in flavor. Epicures even go as far as putting the sauce into egg noodles once they finish eating the lobsters. Noodles fully soaked in the soup taste even more delicious.

入伊面，让伊面充分吸收上汤龙虾的精华，更是一种老饕吃法。

而家常味美的芹菜吊片则是另一道在港澳两地十分受欢迎的新式粤菜。所谓吊片，就是新鲜鱿鱼用竹竿穿起吊住晒干。这道菜在制作时先将鲜鱿鱼斜刀切薄片，浸入清水中，芹菜切段，胡萝卜切条，菠萝切盖挖空。再把鲜鱿鱼放入滚水中，焯至卷起。锅内入油、爆葱、姜、蒜茸，下芹菜、胡萝卜、红辣椒、鱿鱼片，翻炒数下，加盐、味精、姜汁、米酒、勾芡、翻炒，淋油，倒入菠萝内即可。其成菜鲜香味美，滑嫩不腻，令人回味无穷。

港澳两地开放程度高，世界各种文化在此汇聚，其生活方式颇具国际特色，而这一点在饮食上的表现尤为明显。当年英国占据香港以后，大批英法美等外国人来到香港，西餐馆逐渐在香港兴起。西餐入港，一方面为当地带来了更为丰富多彩的菜品和烹饪技巧，一方面也悄悄地改变着香港本土菜和港式粤菜。西餐注重食材的原汁原味，较为倚重调料和味汁，比如名菜红酒煎鲍鱼，就要用到牛油、食盐、红酒、醋、面粉、胡椒粉、葱蒜、红辣椒粉、塔拉根香草等十余种调料。起初香港并没有华人开设的西餐馆，只有一些高级的中国酒楼聘请西厨做西餐，专供欧美人享用。后来，西餐的

Another popular new-style dish dried squid with celery seems more homely. Locals call the "dried squid" "diao pian" meaning to string fresh squids on bamboo poles and hang them to dry. To cook the dish, first, slice the fresh squid at a 45-degree angle and soak them in clear water. Then cut celery into short pieces and carrot into strings. Next, remove the top of pineapple and make it hollowed out. After that, put the fresh squid into boiling water until it rolls up. Then fry scallion, ginger, and bruised garlic and then add celery, carrot, red pepper, and sliced squid to sauté quickly. After that, add salt, MSG, ginger juice, and rice wine until the sauce thickens. After drizzling a little oil, pour the mix into the hollowed pineapple. The cooked dish is savory and smooth, and it is not greasy. It also leaves a long-lasting flavor in the mouths of diners.

Hong Kong and Macao have significantly opened up to the world allowing different cultures to thrive there. The lifestyles of the two cities are decorated with international features as particularly evidenced in their food. During the occupation of Hong Kong by Britain, western restaurants were set up to meet the demands of western customers including British, French, and American people. On one hand, the introduction of Western Cuisine has brought variety into the dish types and cooking methods in Hong Kong. On the other hand, it has changed the region's local cuisine and the Hong Kong-style Guangdong Cuisine. Western Cuisine values the original flavor of food and relies more on seasonings and sauces. An example is the famous Fried Abalone in Red Wine that needs over ten kinds of seasonings including beef tallow, salt, red wine, vinegar, flour, pepper, onion, garlic,

烹饪技术逐渐为在港华人所掌握，国人自己开设的西餐厅便出现了，并且随着人们生活水平的提高，普通老百姓也可以常去西餐厅享用美食了。在这个过程中，以香港厨师为代表的华人厨师不断吸收和创新，一方面保留中餐的传统特色，另一方面又吸取西餐之长，遂创造了一批融汇中西特色的美食，如西法大虾、西法鹅肝、纸包鸡、华洋里脊、铁扒牛肉等。

澳门更早受到西方饮食文化影响，因此其中西合璧的特色更显浓郁。自葡萄牙人占据澳门以后，西方的生活方式便随之入澳，带有明显中西融合特色的澳门饮食文化也逐渐形成。细究澳门饮食中的国际特色，数葡国风情最值一提，其中最为著名的又莫过于马介休。马介休，来自葡语Bacalhau，是鳕鱼经盐腌制而成，为不少澳葡式美食的主要材料。它可以用煎、烧、炒、炸、煮等多种方式烹调，都会令人齿颊留香，回味无穷。澳门不少茶餐厅都有马介休菜式供应，较著名的有西洋焗马介休、薯丝炒马介休、炸马介休球、白焓马介休等。薯丝炒马介休以新鲜的薯仔配以马介休、葡式肉肠等大火炒成，成菜鱼肉鲜美，薯丝香脆味浓。而炸马介休球则可以最充分地释放马介休的肉香，此菜以马介休为主料，然后加上马铃薯泥、洋葱、青椒等碎料拌匀，捏成球，入油锅炸，炸至金黄，装盘即可。其成品金黄诱人，咬

red pepper powder, and tarragon vanilla. At first, there were no western restaurants owned by Chinese in Hong Kong except for a few upscale ones that hired western chefs to cook western food specially for European and American customers. Later, Chinese chefs gradually learned how to cook western food, so they started to open their own restaurants. Meanwhile, as their living standards improved, ordinary people in Hong Kong could afford to go to the western restaurants. In this process, Hong Kong chefs continued to innovate Western food. Building on the traditional characteristics of Chinese Cuisine, the chefs integrated the advantages of Western food and then created a number of dishes combining Chinese and Western characteristics. Some of them included Western-style Prawns, Western-style Foie Gras, Deep-fried Chicken in Tin Foil, Tenderloin cooked in Chinese and Western styles, and Iron Platter Beef.

Given the fact that Macao had an earlier contact with Western culture, its cuisine stood symbolized the combination of Eastern and Western culinary culture. When the Portuguese occupied Macao, they also brought the western way of life with them thereby contributing in integrating two cultures. The most famous example of this cultural integration appears to be bacalhau. Bacalhau, in Portuguese, means "salted cod, "which serves as the main ingredient of multiple Portuguese-Macao dishes. Whether it is fried, burned, deep-fried, boiled or cooked in other ways, it tastes so savory that people who eat it always remember it. Many tea cafes in Macao serve bacalhau-based dishes and the more famous ones include Western Stewed Bacalhau, Stir-fried Bacalhau with Shredded

一口满嘴脆香，让人欲罢不能。除了马介休，澳门典型的葡国菜还有葡国鸡、咖喱蟹、八爪鱼饭等。

如今的香港和澳门，不仅传承着中国饮食的传统，保留着粤菜精髓，融汇着西餐特色，同时也将日式料理、东南亚美食等种种外国菜系纳入自身餐饮体系。在多种元素荟萃的背景下，港澳菜必将不断吸收和创新，融合各方优势，创造出独一无二的饮食文化。

Potatoes, Fried Bacalhau, Bacalhau Cozido. The dish Stir-fried Bacalhau with Shredded Potatoes is cooked by using fresh potatoes, bacalhau, sausage, among other ingredients. After stir-frying, the bacalhau is fresh and delicious while the potato shreds are crisp and flavorful. Another dish, Fried Bacalhau Balls can express bacalhau's taste to the fullest. Bacalhau is the main ingredient of this dish. It is mixed with mashed potatoes, onions, green peppers, and other crumbs and it is stirred well to make meatballs. The meatballs are then fried in a pan until they turn golden brown before they are served on a plate. The cooked dish is golden in color, crispy and tasty in flavor and it is irresistible to anyone. In addition to bacalhau, typical Portuguese dishes in Macao Cuisine also include Portuguese Chicken, Curry Crab Octopus Rice, etc.

Nowadays, given their long history of development, Hong Kong Cuisine and Macao Cuisine accommodate not only Western-style Cuisine, but also many others including Japanese Cuisine, and Southeast Asian Cuisine while preserving the essence of Guangdong Cuisine. In today's world, where different culinary cultures constellate and integrate, Hong Kong and Macao cuisines will continue to absorb new elements and create their own unique stories.

第十章

渊源，台湾的情味

Ten

Taiwanese Cuisine:
Attached Sentiment

　　中华大地幅员辽阔，从河洛地区向东北望去，滔滔的黄河水奔流向渤海，转向东南，在一千余公里外的海域，坐落着宝岛台湾。作为中国的第一大岛，台湾岛物产丰富，海产尤多，在文化的孕育之中，物产变为美食，有了台湾的情味。

　　中华文明是在以河洛为中心的中原地区发展繁荣起来的，而台湾受中华文明的影响发展起来，并成其为一部分，这其中有着深厚的渊源。元朝时首次在台湾设置官署。明清至近代，台湾则多遭磨难。明末荷兰侵占台湾，清初郑成功收复台湾。康熙时，郑克爽归顺清廷，正式定名"台湾"。磨难当中的台湾也因汉族人的移居而得到了更快的发展，饮食文化也随之进步。《马关条约》之后，台湾被日本窃据五十年之久，解放战争之后又与大陆分隔。但这段时间内，台湾的饮食产生了更多的变化，呈现多元的特点，以海味为

The territory of China is vast and huge. Standing on the land of the Heluo area where the Chinese civilization originated and facing the northeast, you can perceive the Yellow River flowing ceaselessly to the Bohai Sea. Turning to the southeast, you can enjoy another picture — the precious island of Taiwan over a thousand kilometres away in the sea. As the largest island in China, Taiwan is rich in natural resources and it has a large variety of seafood. As the culture has developed, these seafood resources have turned into delicacies and now embody the sentiment of Taiwan.

The Chinese civilization has developed and prospered in the Central Plains region ❷ around the Heluo area, and Taiwanese culture has been heavily influenced by and has become a part of Chinese civilization. This has been more evident in the historical process. During the Yuan Dynasty, the government set up a local government office in Taiwan for the first time. However, from the Ming and Qing Dynasties to modern times, Taiwan has undergone significant sufferings. In the late Ming Dynasty, the Netherlands invaded Taiwan, while during the early Qing Dynasty, Zheng Chenggong successfully took back Taiwan from the Netherlands. At the time of Emperor Kangxi, Zheng Keshuang, grandson of Zheng Chenggong, pledged allegiance to the Qing government, and the island was officially named "Taiwan." Taiwan, as it went

❶

Heluo (河洛)means the Yellow River and the Luohe River. The Heluo area is centered on Luoyang city, Henan Province. Ancient Chinese believed that this was the central place of the world, so, according to Chinese history, many dynasties made Luoyang their capital.

❷

The Central Plains (中原) means the middle and lower reaches of the Yellow River.

主的美食更是如此。

台湾美食的多元，与大陆有着千丝万缕的关系。具体来说，传统八大菜系，在台各有体现，比如有悦宾楼擅做鲁菜，有连锁的彭园擅做湘菜，有熟于浙菜的荣荣园，有可烹川菜的骥园等。闽菜、粤菜餐馆在台湾则更多，因为台湾人中有绝大部分是从闽南及广东迁移过去的。民以食为天，食随民而至，台湾闽菜、粤菜之多，使得台湾菜融汇了闽菜和粤菜的烹饪手法和口味特点。因此，台湾的饮食与大陆实际上同宗同源。

台湾菜以海味为主，清淡鲜醇为其特色，同时，又多羹汤，多腌、酱菜。我们先讲一讲台湾本土正式菜肴，一是五味九孔。台湾对鲍鱼俗称九孔，五味九孔一菜，需将鲜鲍鱼过滚水，至肉壳稍将分离，捞出用冷水冲凉。然后剥壳去硬鼻，再将鲍鱼肉翻面放回原壳。此菜五味汁是关键，需用姜、蒜、辣椒、香菜末、葱花及番茄酱、黑醋、酱油、白醋、味精、糖、麻油、酒兑等调配而成。最后将五味汁淋入排列在高丽菜丝之上的九孔即成。这是一道宴席菜，清淡鲜醇尽显无疑，有滋阴清热益精明目的功效。

另一是炒荫豉蚵。此菜是家常菜，制作简单，将牡蛎放盐水稍洗沥干，用开水烫至半熟捞出。然后热锅入油，投豆

through difficult times, has also developed faster because of the migration of Han people from the mainland of China. Meantime, the food culture has developed. After the Qing government was forced to sign the Treaty of Shimonoseki[1] in 1895, Taiwan was illegally occupied by Japan for as long as fifty years. Again, after the War of Liberation (AD1945−1949), the island was separated from the mainland for several decades. However, during this time, Taiwanese Cuisine experienced more changes by embracing diversity especially its seafood cuisine.

The diversity of Taiwanese Cuisine is inextricably linked with the food culture of the mainland. To be specific, the Eight Major Cuisines of China are all reflected in Taiwanese Cuisine. For example, there are Yuebin Restaurant that specializes in Shandong Cuisine, Pengyuan Restaurant that serves good Hunan Cuisine, Rongrong Yuan that specializes in Zhejiang Cuisine, and Jiyuan Restaurant that serves delicious Sichuan Cuisine. Fujian and Guangdong Cuisine restaurants are even more popular in Taiwan given that most Taiwanese migrated from the south of Fujian and Guangdong. For people, food is a necessity and they believe that food goes where people live. There are so many Fujian and Guangdong restaurants in Taiwan. Consequently, Taiwanese Cuisine has integrated the cooking techniques and taste of both of them. Therefore, the diet of Taiwan actually has the same root as that of Mainland China.

The focus of Taiwanese Cuisine is on freshness and mellowness with seafood as the main ingredient. Meanwhile, the cuisine also includes plenty of soup dishes and pickles. Let's

[1]

Treaty of Shimonoseki was an unequal treaty imposed on the Qing government by Japan after the Sino-Japanese war of 1894–1895.

豉、蒜末、蒜苗炒香，下牡蛎、麻油、酒、酱油膏、糖、味精，炒匀即成。此菜虽简单朴素，但鲜美可口。

相比正菜，台湾的小吃更为世人所知。而谈到台湾的小吃，古早味不得不说。古早味，原本是闽南人用来形容古旧味道的一个词，彼时物力有限，人们对食物的处理比较简单，以手工料理为主，现在人们怀念这种味道，故而称之为"古早味"。古早味在台湾的流行，是一种情味，是一种乡情，更是一种文化。

红蟹米糕就是古早味的代表美食。在台湾，红蟹米糕是传统筵席上必有的咸点心，谁家有了喜事，都要送蒸米糕和煮红蛋向亲友道喜。亲友收到米糕，配红蟹烧制，浓香四溢，使四邻都沾到喜气。今天，红蟹米糕仍受欢迎，但十分考验手艺。

这道咸点心，制作材料有糯米、红蟹、猪肉、香菇、虾米、红萝卜、豌豆、葱等。其中豌豆需稍用盐水烫煮，蟹肉需泡在腌肉汁里。随后糯米倒入油锅中翻炒，见全部糯米都已沾上油光，再把酱油沿着锅边倒进，拌炒至颜色均匀。后将糯米置于蒸笼中大火蒸一个小时。时间快到时，热炒锅，先拌炒香菇、虾米和葱花，随后加入猪肉，炒至肉已变色，再把红萝卜、豌豆也加入同炒，撒下胡椒。最后将蒸好的糯

begin by describing how to make the local entrees known as Fresh Abalone in Spicy Sauce. To cook this delicacy, first, boil the abalone until the flesh is slightly separated from its shell, then take it out and wash with cold water. Next, peel the shell and return the abalone flesh back to the shell on the opposite side. It should be noted that spicy sauce, or five-flavored sauce is the key to this dish. They all need to be mixed well with the following ingredients: ginger, garlic, pepper, minced coriander, chopped scallions, tomato sauce, balsamic vinegar, soy sauce, white vinegar, MSG, sugar, sesame oil and cooking wine. Finally, put the abalone on sliced Chinese cabbage, and finish off by pouring the sauce on the abalone. Fresh Abalone in Spicy Sauce is a banquet dish. After eating it, the freshness and mellowness linger on people's tongues for a long time. In addition, according to the traditional Chinese medicine, it has the positive effect of nourishing Yin, getting heat out of the human body, and it is beneficial to the eyes.

The other typical dish is stir-fried oysters with fermented beans. It is a homemade dish and it is simple to make. First, wash the oysters in salt water and drain the water. Next, put it in boiling water until they are half cooked. Then heat the wok with some oil, add fermented beans, bruised garlic, garlic bolt, and then add the oysters, sesame oil, cooking wine, soy sauce, sugar, MSG and stir well. Despite its simplicity, this dish is delicious and mouth-watering.

Compared to entrees, Taiwanese snacks are more popular around the world. As far as Taiwanese snacks are concerned, it is

worthwhile mentioning Guzao due to its taste. Guzao is originally a word used by people of southern Fujian Province to describe an old-fashioned but memorable taste. In the old days, the resources and techniques were limited, so people usually cooked food in a relatively simple way, and everything was handmade. Today people miss it, so they call it the taste of Guzao. The popularity of the taste in Taiwan can be viewed as an affection, a nostalgic feeling and also part of culture.

Rice cake with crab is one of the representative tastes of Guzao. In Taiwan, this is a salty snack that must be part of traditional banquets. No matter which family has a happy event such as a wedding or a new born baby, they are obliged to send steamed rice cakes and boiled red eggs to their relatives and friends. Those who received them would cook the rice cakes with crabs and ensure that the aroma encircling the neighborhood. This brings happiness and lucks to the neighbors. This little snack is still popular till date even though it requires outstanding skills to make.

To make the snack, we need ingredients including glutinous rice, crabs, pork, mushrooms, dried shrimps, carrots, dried peas, and scallions. Firstly, slightly boil the dried peas with salt water, and soak the crab meat to marinate. Then, pour the glutinous rice into the oil pan and stir fry. When all the rice is mixed with oil, pour soy sauce along the side of the pan and stir fry again until all the rice turns dark in color. Then steam the rice for one hour over high heat. Secondly, put oil in the wok and heat just when the rice is going to be cooked. Stir fry mushrooms, dried shrimps, and

米倒入大碗，与已炒好的各种材料混合，红蟹放在最上面再用强火蒸十分钟即可。依古法制作的红蟹米糕，各种食材的味道渗透糯米，糯米筋道弹牙，蟹肉鲜美柔软，蟹黄浓郁细腻，实在是美味至极。

在台湾，虱目鱼是必不可少的食物，干煎成菜，入锅熬粥。虱目鱼粥就是有名的传统小吃。用鱼骨熬的高汤煮生米成粥，吃时将去骨的虱目鱼切成薄片或细丝，与当地风味佐料相伴，入粥稍煮即成。这道小吃里，粥是陪衬，鱼才是主角，其肉质腴美滑嫩，鲜香可口。台湾街道的清晨，匆匆的行人进入小店，吃上一份鱼肠，再吃完一碗虱目鱼粥，如果时间充裕，还可吃鱼皮，喝鱼头汤，提起一天的精神。

关于虱目鱼，还有一则传说，据说郑成功收复台湾行军到台南时，因军饷不够，便派人寻找当地食材充实军粮。一晚，郑成功梦见妈祖，指点他后港有很多鱼，第二天郑成功让士兵到后港捕鱼，果然大丰收。但捕获的鱼郑成功未曾见过，于是用闽南语问是"什么鱼"，旁人误认为这是鱼的名字，于是这种鱼从此就叫"虱目鱼"了，也被称为国姓鱼。

在台湾，面食并不占主流，但小吃大肠蚵仔面线却十分

chopped scallions, then add pork and continue to stir fry until the meat changes color. Thereafter, add carrots and peas and sprinkle a touch of pepper. Finally, put the steamed glutinous rice into a large bowl and mix it with the various cooked ingredients. Place the crabs on top and steam over high heat. Ten minutes later, the snack is done. Rice cake with crab is made in the old way using a variety of ingredients with different flavors interfused with the glutinous rice. This snack is soft, fine and delicious.

In Taiwan, milkfish is an indispensable delicacy. It can be fried as a dish or boiled as congee. The Milkfish Congee is a famous traditional Taiwanese snack. To make it, use fish bones and seasonings to stew fish soup, then add rice into the soup to make congee. When eating, cut the milkfish without bones into thin slices or filaments, add the fish and the various local flavors into the congee, and cook for a while. In this snack, the congee plays a supporting role while the milkfish plays the leading role thanks to its tenderness, smoothness, and deliciousness. In the early morning on the streets of Taiwan, pedestrians in a rush often enter a small restaurant, order a piece of fish intestines, and enjoy a bowl of Milkfish Congee. If they have enough time, they would like to have pieces of milkfish skin and to drink some fish head soup to start off their day.

There is also a legend about the milkfish. It is said that when Zheng Chenggong recovered Taiwan and his troops marched in, he sent soldiers to find local ingredients to enrich the army provisions. One night, he saw the goddess Mazu in his dream. The goddess pointed out that there were a lot of fishes at the port. The

出名。这一小吃最初源自厦门、泉州一代，制作时选用的是金门所产手工面线。在台湾，面线往往是店家自制或采购，制作时大肠事先卤好，面线用高汤熬煮成糊状。面线煮好时，盛入碗中，加蚵仔和大肠，其中鲜蚵仔需稍氽烫，也可加入虾米等提鲜。口感上蚵仔又肥又鲜，大肠油香软韧，面线黏稠，实在美味。从前，大肠蚵仔面线往往出现在路边摊中，各家对蚵仔和大肠各有所爱，放的量或多或少，再配以不同的小料，味道有了些许的差异。对于钟情古早味的人来说，寻自己嗜好的口味，然后手捧烫嘴的面线，临街而食，吃得满头大汗，也是一种畅快。

　　台湾四面环海，台湾人也有因海而生的独特口味，鲨鱼烟便是一例。从前，渔民捕得鲨鱼，要么冰冻，要么制成鱼浆或鱼饵，但都不是妙法。偏偏鲨鱼肉腥臭浓烈，直接烹食难以下咽，因此台湾渔民想出了和大陆西南地区一样的法子——烟熏，于是就有了鲨鱼烟。将新鲜的鲨鱼肉切成大小合适的块，洗净用酒与盐腌制，然后入锅蒸十分钟。再取干锅，锅底铺上糖、茶叶、甘蔗、木屑、稻米，既能上色又能增添香味。锅上铺网架，烟熏至鱼肉金黄即可。最后稍放凉，切片，蘸食。鱼肉、鱼肚、鱼皮、鱼尾，不同部位的鲨鱼烟，或绵软，或弹牙，或劲脆，或多汁，口感也各不相同。如今

next day, Zheng let his soldiers go fishing at the port, and they truly had a bumper harvest. However, Zheng had never seen this kind of fish, so, in Fujian dialect, he asked: "What fish?" Some soldiers mistakenly thought that was the name of the fish. From that time, this fish became known as Milkfish.

In Taiwan, noodles are not the mainstream diet, but a snack named thin noodles with intestines and oysters is very famous. This snack originated from Xiamen and Quanzhou in Fujian Province. Originally, the noodles used for this snack are all handmade from Jinmen. For Taiwanese small restaurants, the noodles are often made or purchased by the owners. Before cooking the noodles, first stew pig intestines in marinade. Then the noodles are boiled in broth until they become paste. When the noodles are well cooked, place them in a bowl and add quick-boiled oysters and intestines. Besides, dried shrimps can also be added to boost both the flavor and freshness. The taste of the oysters is fatty but fresh and the intestines are soft and elastic. The noodles are equally sticky and thick. In the past, thin noodles with intestines and oysters often appeared in the roadside stalls with each stall having different amounts of oysters or intestines and different seasonings. This means the tastes were slightly different too. For those who love the taste of Guzao, having a customized taste, holding a bowl of hot noodles, and eating beside the street until they become sweaty, are ways to be happy and relax.

Taiwan is surrounded by the sea, so Taiwanese people also have unique tastes conditioned by the sea. Smoked shark steak is an example of such a sea-related dish. In the old days,

鲨鱼烟多是机器批量生产，口感远不及以前手工制作，人们对从前的味道也就越发怀念。

在台湾还有两种名声在外的小吃，鼎边趖与蚵仔煎，这两种也常常拿来与福建的鼎边糊和海蛎煎相比。实际上鼎边趖与鼎边糊无论是材料，还是制作，都没有什么差异。如果非要找出区别，大概就是台湾用了"趖"来描述米浆在锅沿上摩挲滑过的情态了吧。蚵仔煎与海蛎煎也是如此，只是福建习惯吃时配以菜头酸，台湾则喜欢淋上甜辣酱、酱油膏或是调配的酱汁。除此之外，菜尾汤、红烧鳗、沙虾肉圆、韭菜柴鱼等也都是颇有名气的海味小吃，往往也能在大陆寻得源头。

除了中华美食的渊源，台湾海味饮食中也有着丰富的国际元素。首先要说的就是日料，天妇罗、寿司、烤秋刀鱼、生鱼片等日料在台湾随处可见，同时，日料在台湾本土化的表现也有不少，比如寿司要配以酱油膏，生鱼片厚度比日本厚得多，进餐礼仪也没有日本严肃等。除了日料，法、意、泰、印等国的料理在台湾也都可以见到，其中有追求正宗者，如意料义玛卡多餐厅所做的维多利亚女王炖饭，虾味浓郁，酸咸可口；又如 Thai & Thai 餐厅，烹制泰国海鲜菜肴，专程从泰国空运食材。此外，有的国外料理餐厅则做了一些本

once fishermen caught a shark, they either froze it or made it into fish paste or baits. However, these were not good methods of preserving the fish. Shark meat is so fishy and foul that it is difficult to swallow even if it is cooked. Therefore, like inhabitants of the southwestern part of the mainland, Taiwanese fishermen came up with the idea to smoke the fish thereby creating the dish smoked shark steak. To make the dish, first cut the fresh shark meat into pieces of proper sizes. Wash and marinate it with cooking wine and salt, then steam it in a pot for ten minutes. Then take a griddle pan, and put sugar, tea, sugar cane, sawdust and rice in the bottom of the pan, which would help deepen the color of the fish and add flavor to the dish. Next put a wire frame over the griddle and smoke the fish until it becomes golden. Finally, allow it to cool down slightly before slicing it and serving it with sauce. Now, you can dip to eat. Different parts of the shark, such as fish meat, fish maw, fish skin and fish tail, have different tastes. They are soft, elastic, crisp or juicy. Today, most smoked shark steak dishes are mass-produced by machines, so they are far less tasty than handmade dishes. As a result, people are more and more nostalgic for the old taste.

There are another two famous snacks in Taiwan called Dingbiansuo and Ou-a-jian, which are often compared with Fujian's rice paste in shrimp soup and oyster omelette. In fact, there are almost no differences between Dingbiansuo and rice paste in shrimp soup in terms of ingredients and cooking methods. If you are still interested in the differences, then it is worthy to note that Taiwan used the character " 趖 " to describe

the process of rice water sliding down from the top edge of the wok to the bottom. And this is also similar for Ou-a-jian and oyster omelette except that Fujian people are accustomed to enjoying oyster omelette with sweet and sour pickled moolis, while Taiwan people prefer topping with sweet chili sauce, thick soy sauce, or blended sauce. Besides, seafood snacks like mixed vegetable soup, braised eel, prawn meatballs and Chinese chive and dried cod are also well-known. Their origins can often be traced to mainland China.

In addition to the origin from Chinese Cuisine, Taiwanese seafood dishes also boast a rich variety of international elements. The first one is Japanese food. Typical examples like tempura, sushi, grilled saury, and sashimi can be seen everywhere in Taiwan. At the same time, there are many localized versions of Japanese food in Taiwan. For example, sushi is often served with a thick soy sauce and there are discrepancies between the thickness of the sashimi of Taiwan and that of Japan. Furthermore, the dining etiquette is not as serious as that in Japan either. There are also a lot of French food, Italian food, Thai food, Indian food, etc. in Taiwan. The Italian restaurant IL MERCATO serves the dish Queen Victoria Risotto that has a rich shrimp flavor and it tastes sour and delicious. Another example is Thai & Thai that serves Thai seafood dishes with ingredients specially transported from Thailand. Besides, some foreign restaurants have also adapted their cooking methods and sauces to cater for Taiwanese preferences.

土化，在口味上更符合台湾人喜好，比如烹饪方式、所用酱汁等，都稍有调整。

无论是精致的海鲜菜肴，还是古早的海味小吃，寻根溯源，往往都能找到往昔的痕迹。台湾丰饶的物产孕育在海洋的气息之中，化为美食，留香于唇齿之间，余情于脑海之中，点点滴滴，是传承的文化，是萦绕的乡情，也是台湾的情味，是与大陆剪不断的渊源。

Whether in exquisite seafood dishes or seafood snacks of the Guzao taste, there are consistent traces of the past. The rich and fertile resources are bred in the ocean, turned into food, scented between the lips and teeth and systematically engraved in our minds. All of these are the manifestation of the inherited culture, a lingering nostalgia, a sentiment of Taiwan, and an inseparable attachment of Taiwan to mainland China.

古代

〔东周〕孔丘编，王世舜、王翠叶译注：《尚书·禹贡》，中华书局，2012 年。

〔战国〕屈原、〔战国〕宋玉：《楚辞》，湖北辞书出版社，2017 年。

〔西汉〕戴圣：《礼记·王制》，中州古籍出版社，2016 年。

〔西汉〕刘安：《淮南子·精神篇》，岳麓书社，2015 年。

〔西汉〕刘歆、〔西汉〕刘向：《皇帝内经》，北京燕山出版社，2010 年。

〔西汉〕司马迁：《史记》，中华书局，2008 年。

〔东汉〕赵晔：《吴越春秋·王僚使公子光传》，江苏古籍出版社，1986 年。

〔晋〕嵇含：《南方草木状》，广东科技出版社，2009 年。

〔南朝宋〕范晔：《后汉书·东夷传》，延边人民出版社，1995 年。

〔北魏〕贾思勰：《齐民要术·卷七》，团结出版社，1996 年。

〔唐〕段成式：《酉阳杂俎》，上海古籍出版社，2012 年。

〔唐〕刘恂撰，鲁迅校勘：《岭表录异》，广东人民出版社，1983 年。

〔唐〕徐坚等著：《初学记》，中华书局，1980 年。

〔宋〕范成大撰，严沛校注：《桂海虞衡志校注》，广西人民出版社，1986 年。

〔宋〕赵汝适：《诸蕃志·海南》，中华书局，1956 年。

〔宋〕周去非：《岭外代答》，中国书店出版社，2018 年。

〔宋〕朱肱：《北山酒经》，上海书店出版社，2016 年。

〔元〕高德基：《平江记事》，四库馆，1868 年。

〔明〕冯梦龙：《东周列国志》，华夏出版社，1994 年。

〔明〕兰陵笑笑生：《金瓶梅》，上海古籍出版社，2005 年。

〔明〕李时珍：《本草纲目》，北京联合出版公司，2015 年。

〔明〕宋应星：《天工开物》，甘肃文化出版社，2003 年。

〔明〕唐胄：《琼台志》，海南出版社，2006 年。

〔明〕王鏊：《姑苏志》，四库馆，1868 年。

〔清〕毕懋第修：《威海卫志》，天津古籍出版社，2013 年。

〔清〕蔡麟祥修、〔清〕林豪纂:《澎湖厅志》,1956 年。

〔清〕二石生:《十洲春语》,上海申报馆,1877 年。

〔清〕怀荫布主修:《泉州府志》,泉州市地方志办公室影印,1985 年。

〔清〕季麒光等:《台湾杂记 台湾纪略》,商务印书馆,1959 年。

〔清〕蒋毓英:《台湾府志》,中华书局,1985 年。

〔清〕乔弘德纂修:《安东县志》,上海社会科学院出版社,1989 年。

〔清〕屈大均:《广东新语》,中华书局香港分局,1974 年。

〔清〕沈定均修,〔清〕吴联薰增纂:《漳州府志》,中华书局,2011 年。

〔清〕王守恂:《天津县新志》,江苏古籍出版社,2010 年。

〔清〕王守恂:《天津政俗沿革记》,1938 年。

〔清〕谢道承、〔清〕郝玉麟:《福建通志》,四库馆,1868 年。

〔清〕印光任、〔清〕张汝霖:《澳门记略》,广东高等教育出版社,1988 年。

〔清〕袁枚:《随园食单》,中华书局,2010 年。

〔清〕张庆长:《黎岐纪闻》,上海书店,1994 年。

〔清〕张焘:《津门杂记》,天津古籍出版社,1986 年。

〔清〕张心泰:《粤游小识》,1900 年。

[清] 郑杰：《闽中录》，海风出版社，2001 年。

[清] 周亮工：《闽小纪》，福建人民出版社，1985 年。

现代

[英] 艾伦·艾萨克斯：《麦克米伦百科全书》，郭建中译，浙江人民出版社，2002 年。

蔡澜：《寻味南半球》，青岛出版社，2018 年。

蔡澜：《寻味欧洲》，青岛出版社，2018 年。

蔡澜：《寻味日韩》，青岛出版社，2018 年。

陈铭枢：《海南岛志》，上海神州国光社，1933 年。

陈文：《加拿大粤菜》，广东科技出版社，1999 年。

陈宗兴、顾文选、陶斯亮主编：《中国城市大典·第一卷》，人民日报出版社，2005 年。

澄海县地方志编纂委员会：《澄海县志》，广东人民出版社，1992 年。

崔林涛等主编：《中国历史文化名城大辞典》，中国人事出版社，1995 年。

大连市史志办公室编：《大连市志：民俗志》，方志出版社，

海错食单
The Stories of
the Chinese Seafood

196 • 197

2004 年。

戴愚庵:《沽水旧闻》,天津古籍出版社,1986 年。

丁宜曾:《农圃便览》,中华书局,1957 年。

杜福祥、谢帼明主编:《中国名食百科》,山西人民出版社,1988 年。

《广州百科全书》编委会编:《广州百科全书》,中国大百科全书出版社,1994 年。

国家海洋局科技司:《海洋大辞典》,辽宁人民出版社,1998 年。

海南地方文献丛书编纂委员会汇纂:《康熙万州志·道光万州志》,海南出版社,2004 年。

海南省琼山市地方志编纂委员会编:《琼山县志》,中华书局,1999 年。

韩欣主编:《美食中国·下卷》,北京科文图书业信息技术有限公司,2007 年。

河本方主编:《中华民国知识词典》,中国国际广播出版社,1992 年。

胡国年主编:《中华民俗大全(澳门卷)》,《中华民俗大全·澳门卷》编辑委员会出版,2003 年。

胡乔木主编:《中国大百科全书》,中国大百科全书出版社,1993 年。

华英杰、吴英敏等编：《中华膳海》，哈尔滨出版社，1998 年。

《黄河文化百科全书》编纂委员会编：《黄河文化百科全书》，四川辞书出版社，2000 年。

季鸿崑、李维冰、马健鹰：《中国饮食文化史·长江下游地区卷》，中国轻工业出版社，2013 年。

季鸿崑：《烹饪学基本原理》，中国轻工业出版社，2016 年。

焦桐：《台湾味道》，生活·读书·新知三联书店，2011 年。

金开诚：《诗经》，中华书局，1980 年。

[俄] 康斯坦丁·谢平：《美食中国》，人民文学出版社，2018 年。

黎小江、莫世祥主编：《澳门大辞典》，广州出版社，1999 年。

李少兵：《民国时期的西式风俗文化》，北京师范大学出版社，1994 年。

李夕聪、邓志科、周德庆编：《环球海味之旅》，中国海洋大学出版社，2018 年。

李元春：《台湾志略》，台湾银行，1958 年。

李治亭主编：《爱新觉罗家族全书·五》，吉林人民出版社，1997 年。

连横：《台湾通史》，人民出版社，2011 年。

辽宁省盘山县地方志编纂委员会：《盘山县志》，沈阳出版社，

1996 年。

林祥庚:《福建民俗》,福建师范大学出版社,1989 年。

林正秋等编:《中国饮食大辞典》,浙江大学出版社,1991 年。

林众主编:《中华旅游通典》,社会科学文献出版社,2004 年。

刘乾先主编:《中华文明实录》,黑龙江人民出版社,2002 年。

卢德平主编:《中华文明大辞典》,海洋出版社,1992 年。

卢乐山:《中国女性百科全书·社会生活卷》,东北大学出版社,1995 年。

陆国俊:《美洲华侨史话》,商务印书馆,1997 年。

陆军主编:《中国江苏名菜大典·上册》,江苏科学技术出版社,2010 年。

聂作平编著:《中国神话传说》,天津教育出版社,2007 年。

牛汝辰:《中国水名词典》,哈尔滨地图出版社,1995 年。

潘倩菲:《实用中国风俗辞典》,上海辞书出版社,2013 年。

彭勇:《明史》,人民出版社,2019 年。

强振涛:《中国闽菜》,福建人民出版社,2018 年。

乔德秀:《南金乡土志》,哈尔滨出版社,2003 年。

曲金良编著:《海南民俗》,甘肃人民出版社,2008 年。

上海书店出版社编:《中国地方志集成·海南府县志辑》,上海书店出版社,2014 年。

石泉长主编：《中华百科要览》，辽宁人民出版社，1993 年。

史为乐编：《中国历史地名大辞典》，中国社会科学出版社，2005 年。

史仲文、胡晓林主编：《中华文化习俗词典》，中国国际广播出版社，1998 年。

台湾省文献委员会编：《台湾省通志稿》，台湾省文献委员会出版，1957 年。

天津市档案馆编：《天津商会档案汇编》，天津人民出版社，1997 年。

天津市地方志编修委员会：《天津通志》，天津社会科学院出版社，1998 年。

王者悦主编：《中国药膳大辞典》，大连出版社，1992 年。

天津市地方志编修委员会编著：《天津旧志点校》，南开大学出版社，2001 年。

天津市旅游局、天津科学技术出版社合编：《天津指南》，天津科学技术出版社，1983 年。

田曙岚：《海南岛旅行记》，中华书局，1936 年。

铁木尔·达瓦买提编：《中国少数民族文化大词典》，民族出版社，1999 年。

汪少云：《中华药膳·上卷》，天津古籍出版社，2007 年。

王克信:《天津民俗》,南开文艺编辑部,1991年。

王树楠、吴廷燮、金毓黻等纂:《奉天通志》,辽海出版社,2003年。

[美]威廉·C.亨特:《广州番鬼录》,冯树铁、沈正邦译,广东人民出版社,1993年。

吴海林、李延沛编:《中国历史人物辞典》,黑龙江人民出版社,1986年。

吴瀛涛:《民俗台湾》,众文图书公司印行,1987年。

武清区地方史志编修委员会编著:《武清县志》,天津社会科学院出版社,2004年。

冼剑民、周智武:《中国饮食文化史·东南地区卷》,中国轻工业出版社,2013年。

徐葆光:《福建通志台湾府》,台湾银行,1960年。

徐公武:《海南岛》,新中国出版社,1948年。

徐海荣:《中国美食大典》,浙江大学出版社,1992年。

徐杰舜:《汉族风俗史·第一卷》,学林出版社,2004年。

徐杰舜:《汉族风俗史·第五卷》,学林出版社,2004年。

徐珂编:《清稗类钞》,中华书局,2010年。

徐兴海、胡付照:《中国饮食思想史》,东南大学出版社,2015年。

徐肇琼:《天津皇会考 天津皇会考纪 津门纪略》,天津古籍出版社,1988年。

臧维熙:《中国旅游文化大辞典》,上海古籍出版社,2000年。

张岱年主编:《中国文史百科》,浙江人民出版社,1998年。

张定亚主编:《简明中外民俗词典》,陕西人民出版社,1992年。

张静如、刘志强:《北洋军阀统治时期中国社会之变迁》,中国人民大学出版社,1992年。

张仲编著,来新夏主编:《天津早年的衣食住行》,天津古籍出版社,2004年。

赵冬:《走遍美洲》,山西人民出版社,2003年。

赵利民主编:《中国传统保健菜谱》,天津古籍出版社,2009年。

《哲学大辞典》编辑委员会:《哲学大辞典》,上海辞书出版社,1985年。

郑友军编:《新版调味品配方》,中国轻工业出版社,2002年。

《中国城市发展全书》编委会编:《中国城市发展全书》,中国统计出版社,2004年。

中华文化通志编委会:《中华文化通志·饮食志》,上海人民出版社,1998年。

周剑箫:《江南才子唐伯虎》,中信出版社,2018年。

周宪文等编:《台湾文献史料丛刊》,人民日报出版社,2009年。

周元文：《重修台湾府志》，台湾文献馆，1930 年。

朱道清编：《中国水系词典》，青岛出版社，2007 年。

朱振藩：《味兼南北》，生活·读书·新知三联书店，2016 年。

从西周到今天

传承三千余年

自辽东至海南

丈量三万多千米海岸线

寻味中国人舌尖上的十二时辰

历史典故、民间传说、风土人情荟萃盛宴